CW01369041

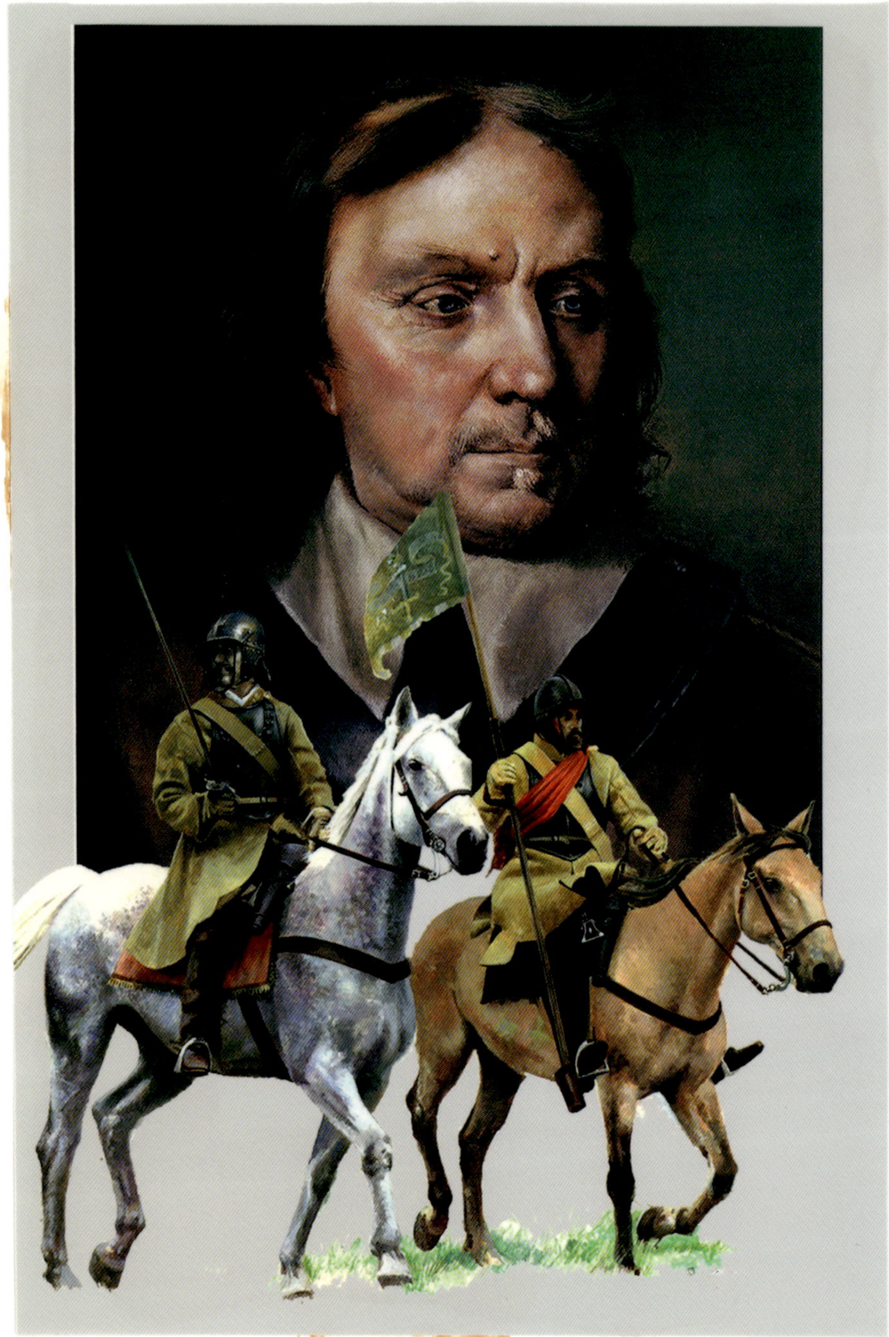

Cromwell's Soldiers

The Moulding of The New Model Army
1644-1645

by
Barry Denton MBE, FRHistS

Cover and Illustrations
by
Chris Collingwood

© 2004 by Sandra Lubinsky Ossanna. All rights reserved.

To Arrow
With Love, Quiver.

No part of this book may be reproduced, stored in a retrieval system, or transmitted by any means, electronic, mechanical, photocopying, recording, or otherwise, without written permission from the copyright owner.

Published by Denton Dare Publishing Company -
www.dentondare.co.uk

ISBN:0 -9762542-9- 8 (Dust Jacket)
ISBN:0-9762542-7-1 (E-Book)

Printed in the United States of America
and England

This book is printed on acid free paper.

PREFACE

To meet Barry Denton was a challenge. For to meet Barry Denton was to meet an impossibly indomitable spirit. He was irrepressibly curious, committed and fun. His annual reports to the Cromwell Association AGM, commenting on his work as press officer, revealed his insatiable appetite to ensure that the name of Cromwell be kept in the public domain, and they were written with an impish humour that caused what could be rather earnest gatherings to lighten up. He simply refused to allow the circumstances of his life, which largely confined him to his home to get in the way of research. And it seems never to have occurred to him that it was severely impractical for a quadriplegic to make retrievable notes out of which he could compose works of history that could be enjoyed by others who shared his passion for the civil wars. The works he had written by 1996 brought him recognition in the form of election to a Fellowship of the Royal Historical Society. When the Council elected him, only I knew of the circumstances of his life. He was elected strictly on merit. It was a unique triumph, and it was rooted not in self-belief, but in an unselfconscious determination to forget his personal circumstances and to get on with answering the questions he wanted to have answered. That passion arose both from an intrinsic fascination with the period, and from a passionate respect for what men like Cromwell stood for: social justice, religious liberty, political rights.

To meet Barry Denton was a challenge. Disabled people are not supposed to be so full of joie de vivre. His writings are like the man: bubbling away with enthusiasm, passion, good humour. His writings resemble the expeditions of an inveterate fell-walker: the purposeful stride, the relish in the steep ascents, and the regular stops to take in, admire, soak up the vistas.

Cromwell's Soldiers is his masterpiece, the biggest, boldest, most ambitious of all his works. It is wonderful that he lived to complete it, sad that he did not live to see it published.

Barry is not the first to examine the origins of the New Model Army. And while his reading was wide-ranging and varied (official records of the Parliament and its committees, the semi-official compilations of those who were clerks to the two Houses and to the Army, a full range of the weekly newspapers of the time, an excellent cross-section of the pamphlets, and an exact knowledge of the memoirs of many of those involved) it was confined to printed sources. This is a limitation, although not a disfiguring one, I think. It is largely made up for the attentiveness of Barry's reading, his seeing so much more in the sources than others have seen.

One small example of this is his acute observation on the way the conscription act of early 1642 excluded from forced service precisely the people that Henry Ireton was later to identify at Putney as having a 'fixed permanent interest' in the kingdom.

As well as having a very keen eye, Barry had a very sharp ear. One of the most attractive things about this book is the way that he knows when to summarise and gloss his sources and when to offer the reader good meaty chunks of seventeenthcentury prose. His judgment on this seems to me to be well-nigh impeccable, and the result is the reader gets a real sense for the vigour and passion of debate. Absolutely typical of this is Barry's decision to include as appendix 4 a rare and little-known tract entitled A narration of the expedition to Taunton which, as much by its tone as by its content, gives us a wonderful feel for the exhilarating early months of the New Model. Barry's detective work on the anonymous author of this tract is typical of that attentive shrewdness to which I have already alluded.

This book is in essence first-rate story-telling. Barry displays a gift for narrative that rarely deserts him. Some of the more analytical passages – such as the social analysis at the beginning of chapter 3 – are seriously open to challenge; and I suppose in truth it must be admitted that Barry is sometimes a little too willing to take memoirs written many years later at face value, as representing the views men held in 1645 rather than views coloured by the wisdom of hindsight. His reliance on the autobiography of Richard Baxter, a puritan minister who became deeply antipathetic to Cromwell and those close to him, is a case in point. They are small blemishes on a book that is saturated with Barry's enthusiasm, energy, keenness of observation, and which remind us of the twinkle that was ever in the eye. It has been a great pleasure to undertake as his surrogate the kind of tidying up that any author undertakes between completing a draft of his or her book and handing it over to the publisher.

Responsibility for the book's appearance lies first and foremost with Sandra Dare Ossanna, a good friend to Barry. She was determined that this would be a memorial to him. She had the imagination and flair to assemble the team that has taken the disks that Barry left and prepared them for publication. She commissioned the illustrations. She invited me (who had worked with Barry in the Cromwell Association, whose President I had the honour to be from 1989 to 1999) to write this preface and to check the typescript.

It was vital that we did not change anything that Barry had wanted to say, but what he left could not be published as it stood. Many of the footnotes were incomplete or were in abbreviated form; there were many typos and tiny inaccuracies of spelling and punctuation such as inhabit any first draft I have ever seen. There were some sentences whose meaning was hard to establish. All the technical defects were corrected by Victoria Gregory, who had just completed a PhD on seventeenth-century history at Cambridge. The work of tweaking the style so as to remove the imprecisions without changing the meaning was undertaken by me. At the very beginning of chapter 4, I have added three lines of text. For there Barry makes reference to the Treaty of Uxbridge without ever explaining what it was. Otherwise, I have simply undertaken the role of a copy-editor with a light touch. Every change is intended to clarify not to change the meaning, and I have resisted every temptation to go beyond that brief. It has been for me – as I know for all those involved in this project – a labour of love.

JOHN MORRILL
Professor of British and Irish History
University of Cambridge
February 2003

"To all of Barry's friends and colleagues whom Barry held so dear in his heart."

INTRODUCTION

What was the New Model Army? To the man in the street it possibly conjures up a picture of a soldier in triple barred helmet, dour and stern of temper, the grimfaced puritan compared to the archetypal laughing cavalier. This is indeed the guise popularised by many works of fiction, but history tells us much more than this grey symbolic vision of the Cromwellian trooper can portray. To the military historian the New Model was the first English national, perhaps even nationalist, attempt to organise a regular army. To the political historian the same army can become the radical centre for change, the stage upon which conservatism, libertarianism and primitive socialism met head on. The history of literature also presses its attention on the words of soldiers who lived and died three centuries ago, whether with the radicalism of John Lilburne, the speeches of Oliver Cromwell, the beauty in the poetry of John Milton, or the tortured prose of one time soldier John Bunyan. Whichever route the historian takes with the New Model Army, the march is full of the illumination which brings history alive. The ethos of English, and naturally Scottish, puritanism might well be dour, the purity of conscience relating, and taking to its mantle, a darker side to the soul of its military standard bearer. Yet in this army was to be found keys of cold reality to a nation's future, the tool by which the restrictive lock of posterity could be released.

Historical interest in the New Model Army has grown in recent years mainly through a small number of important works. Interest has been primarily in the political debates of 1647-1649, reintroduced by Sir Charles Firth into the general history of the period. Firth and slightly earlier his mentor, S. R. Gardiner, were in the main responsible for the re-analysis of sectarian elements in the army, and through his studies of the Clarke Manuscripts (Worcester College Library, Oxford), Firth's reintroduction into the debate of the so-called Levellers and Anabaptist factions in political society and primarily the military. This carefully conservative, yet essential, grounding in the radical elements of the army by Firth were eagerly grasped by Marxist historians like Christopher Hill, A. L. Morton and H. N. Brailsford who grafted the flesh of John Lilburne and Edward Sexby onto the skeletal frame of their doctrine, giving it a fresh perspective in its own time, and perhaps in no small way supplying a foundation missing from the politics of the left in this country. It is interesting that the characterization of left-wing theory, in the mouths of Lilburne and his fellow travellers, as depicted by Hill and his colleagues, has emerged almost uncannily in the turmoil of modern day Eastern Europe in the actions of its people. Perhaps it is no bad analogy to compare the social liberalism of John Lilburne with the later cause seen in the shape of Vaclav Havel, and to conclude that the former was simply born before, or out of sympathy, with the time in which he lived. This work of Christopher Hill has been important, even invaluable, in re-establishing the sects, as opposed to sectarianism, to the seventeenth century, albeit it is, in the main, very much a chicken and egg scenario.

This present volume does not reach the 1647 period of Levellerism, indeed it was only ever my intention to carry this study up to Naseby. My aim has been to examine why the New Model came into being, the actual act of moulding the first national regular army.

A further group of historians, inhabiting the more liberal mainstream ground, but willing to accept or utilize the cases hypothesized by Hill and his contemporaries, and also the classical works of Firth and his, come in the guise of Austin Woolrych, Gerald Aylmer, John Morrill, Ivan Roots and Ian Gentles.

The spur which caused the resurgence of interest in the New Model, was the work of Mark Kishlansky, and primarily his 1979 book The Rise of the New Model Army (Cambridge, 1979). Kishlansky's revisionist view of the politics surrounding the first two years of the army's rise to power encouraged others to examine its role. In The Rise of the New Model Army, Kishlansky attempted to prove a hypothesis that the army was less radical, less sectarian, or to be fair, if radical and sectarian, to limit the affects this had on events in future years. Fundamentally, Kishlansky argues that the New Model was a reorganization of previous armies from the early years of the civil war. This point is difficult to oppose in purely military terms, yet simply by examining the events of 1644-45 the first two points are thrown back into debate. Kishlansky for example, neatly sidesteps the issues raised by the Covenant, Self Denial and the commission under which the New Model Army would operate, simply by referring to the decisions by Parliament as 'votes' without addressing the significance of their formal adoption as ordinances.

Other historians have responded to the position taken by Mark Kishlansky throughout the 1980s, primarily by studies devoted to the constitutional debates of 1647. In the catalogue of works, one is of particular interest to the study of the New Model in that period. In 1987 Austin Woolrych produced his invaluable account of the army crisis of 1647, Soldiers and Statesmen, which dealt with the reaction of the members and elements of the New Model to the questions set by their own employment, and the consequential questions set by the quest for a national settlement. Woolrych's book, although treading carefully along the path of orthodox history, and acknowledging the better parts of the Kishlansky revision, clarifies much of the general understanding of the army's search for a place in society after the war; its 'Promised Land' perhaps, where the soldiers would see the fruits of their labours. This trauma in the army, however, was two years after the raising of the New Model, and represents a condition resulting from two years of political change, albeit change which had been brought about by that very same army.

This present work covers a period of roughly nine months, with only two being during actual field service of the New Model. It grew from a need to identify the basis from which the army would emerge as a political unit in the spring of 1647. My earlier works Naseby Fight 1645 (Partizan Press, 1991) and Crisis in the Army 1647 (Partizan Press 1985), had seen the New Model in battle array in the fields of my native Northamptonshire, and divided in the political arena two years later. Yet the question remained: did the actual 'moulding' of the New Model Army allow, even encourage, the later events to progress to their conclusions? Mark Kishlansky had made the radicalism of the army secondary to its birth. The question that remained for me, however, was not whether the New Model a hotbed of revolt, but how did the military babe grow in the womb and what characteristics did it gain from its parents. I felt that if these questions to the political gestation of the embryonic army could be answered, these open theories as to whether the later politicising of the forces was indeed grafted onto the army, or was, as I suspected, a genetic influence dating from its conception, would be resolved.

The whole point of raising a new army, appeared to rest in two fine concepts of law: the Solemn League and Covenant which laid down the principle of a Anglo-Scottish alliance and church state in presbyterian mould, and the Self-Denying Ordinance, which removed the members of the peerage from military service. I quickly found that where in 1647 it was easy to divide the lower ranks into political, shall we say, stables, the divisions of 1644-45 relied much more strictly upon an openly religious, or politico-religious line in its radicalism. It is dangerous moreover, to begin to study the moulding of the army by placing emphasis on political and religious points in isolation to each other. Whereas much to do with the question of Self Denial can be seen as political in context, this is equalized by the religious questions set by the Covenant. Mark Kishlansky had found that the New Model had been a continuation of previous armies simply reorganized, but in over simplifying this, he had ignored to a large extent the effects Self Denial and the Covenant had upon this reorganization. I therefore concluded early in the research, that to understand the moulding of the army, this new study must examine and understand the notion understood by our forebears that for every occurrence there must be a reason, whether it is governed by God or man. If this was the understanding by which the New Model was given life in 1645, then only by examining this doctrine in the Covenant could the author and reader find a satisfactory solution to the future actions of this army.

The question of religion within the army however, is related within this volume only to its effect on political judgement, the work of Chaplains within the army already having found a champion in the work of Dr. Anne Laurence. Dr. Laurence's Parliamentary Army Chaplains 1642-1651 (Woodbridge, 1990), already providing an excellent grounding in the finer points of the influences of Hugh Peter, John Saltmarsh, and their brethren.

Likewise, no full regimental histories will be found in this volume. For those, see the original *Regimental History of Cromwell's Army*, edited by C. H. Firth and

G. Davies, 2vols (Oxford 1940), and my own ongoing works for the Partizan Press. Yet in this present volume will be found more detailed assessments of how officers were chosen, and analytical details between the official officer lists.

The final duty in an introduction is to thank those who ease the burden of research and writing of the volume, a pleasure in fact when those mentioned are so often, or become, good friends. I would therefore like to thank Austin Woolrych, Anne Laurence, Ian Gentles and Gerald Aylmer for so generously lending me books from their respective libraries, and generally offering their friendship and encouragement. To Neil Chippendale for obtaining access for me to many old volumes and numerous tracts. To Anne West for assistance with the Calendar of State Papers Domestic series. And lastly, my parents Irene and Ray Denton, for so much physical support during all stages of the work.

ACKNOWLEDGEMENTS

The British Library
The Bodleian Library, Oxford
The Public Record Office, London
Dr. Williams's Library, London
Worcester College Library, Oxford
Somerset County Record Office Northamptonshire Record Office Leicestershire
Record Office Oxfordshire Record Office Wiltshire Record Office

Acknowledgement is also due to the following individuals -
Dr Derek Massarella (Tokyo), Dr Jonathan Scott, Professor Ivan Roots, Major A. J. Flint, Mr Stuart Peachey, Mr Alan Turton, Professor John S. Morrill.

TABLE OF CONTENTS

CHAPTER I
DIVISIONS AND DIVERSIONS 1

CHAPTER II
TOWARDS SELF-DENIAL...16

CHAPTER III
A NEW MODEL FOR THE ARMY..30

CHAPTER IV
THE TAKING OF THE COVENANT 40

CHAPTER V
GENERAL FAIRFAX 54

CHAPTER VI
THE BREAKING OF THE COVENANT 70

CHAPTER VII
AN ENGLISH ARMY..81

CHAPTER VIII
THE CROMWELL FACTOR...101

CHAPTER IX
THE PARTING OF THE ARMY..112

CHAPTER X
OXFORD AND LEICESTER 126

CHAPTER XI
THE ROAD TO NASEBY FIGHT..139

CONCLUSION 152

APPENDIX 1
THE OFFICER LISTS..157

APPENDIX 2
THE MARCHING OF THE ARMY 168

APPENDIX 3
COATS AND COLOURS...172

APPENDIX 4
A TRACT FROM TAUNTON..175

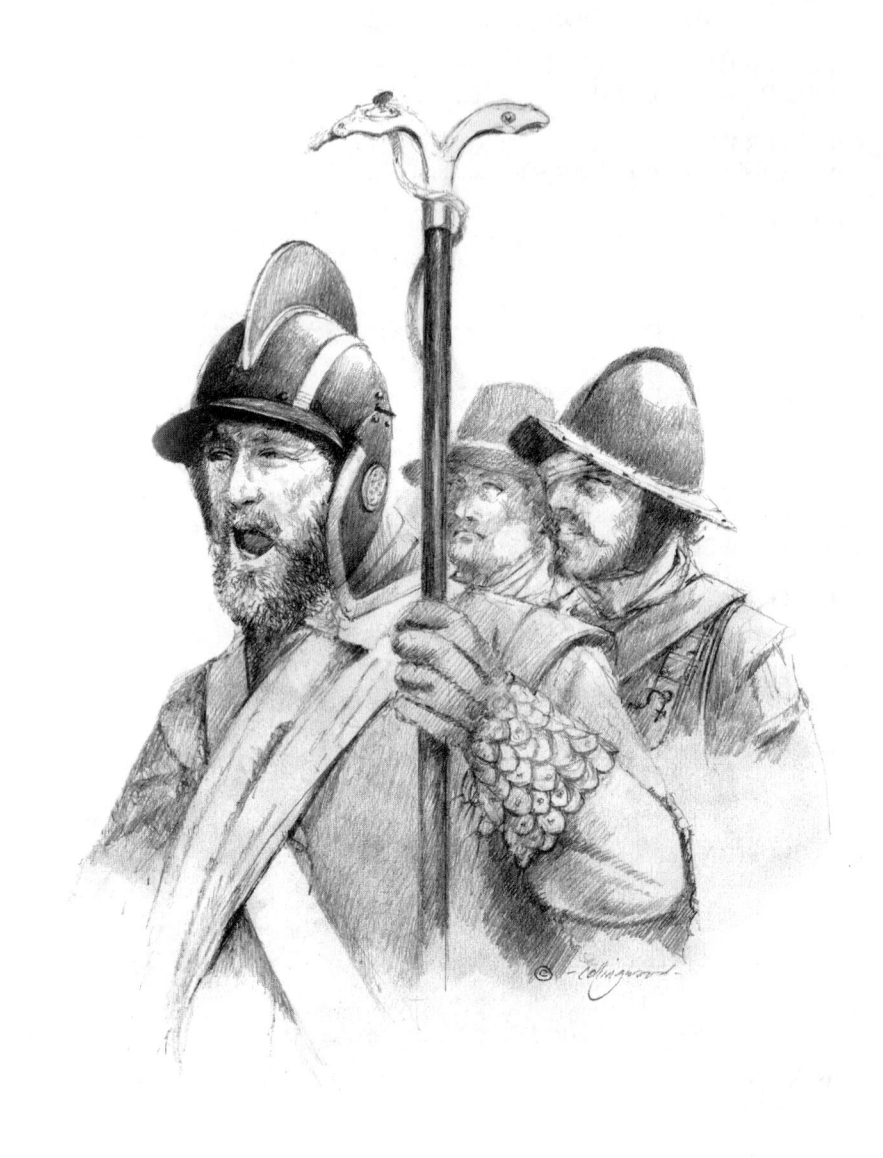

CHAPTER I
DIVISIONS AND DIVERSIONS

It had been a mixed year for the military fortunes of Parliament, their early victories at Cheriton and Marston Moor having been largely reversed at Cropredy Bridge and Lostwithiel. The King's forces were far from being at their lowest ebb despite the intervention of Scottish Covenant troops on the side of the Parliamentary cause, and indeed it was becoming advantageous to the royalists to extend the military conflict to reap political strength in country areas, or even in the City of London itself.

Was this the background to the thoughts in the officers' minds as in October 1644 the smoke of battle cleared from the outskirts of Newbury? For the first time in the war, the army of the Eastern Association commanded by the Earl of Manchester had joined with the Southern and Western forces under Sir William Waller, and the re-equipped army of the Earl of Essex. It should be stressed that the morale of Essex's men was low following their surrender at Lostwithiel, and the Earl himself was absent through illness.[1] The strategy of the battle had been utter chaos. Each commander of horse and foot had fought a fierce engagement, often very successfully, yet in the stance of almost comic opera, none of the commanders had fought at the same time. Phillip Skippon commanding the old foot from Essex's army and the London Trained Bands, had led his 'brave boys' through fire and lead, but without sufficient cavalry to make the strength of his forces tell. Oliver Cromwell who was possibly the finest cavalry officer the war had produced, had failed to make his attack at the right time, due to the difficult ground conditions for horse around the battle area. Waller's forces had followed their officers in the same manner towards Speen on the outskirts of Newbury, but without a co-ordinated attack, they could not inflict the crushing defeat on the King that the opportunity had provided. Manchester himself held his Eastern Association infantry back for most of the day, his hesitation causing the assault to be disorganized, and then ordering the attack when it could achieve no result.

That the Earl of Manchester was reluctant to fight was further emphasised when he failed to support Cromwell's horse in their pursuit of a retreating royalist army at the days end, and then allowed the King to rescue his artillery and supplies from nearby Donnington Castle.[2]

Much of the ineffectiveness of the battle of Newbury could be traced to the whole structure of military command laid down by Parliament in the previous two years. It is quite reasonable, if unpalatable to traditionalist thinking, to describe the civil war as the first foray in an internal capitalist conflict. Certainly the Parliamentary command structure of Associations was based as much on regional economics as religion or the old landed classes. The Midland Association for example was the supply area for horseflesh and shoes from Leicester or Northampton, and arms from the West Midlands. The Eastern Association held the east coast ports, powder magazines, and farmland. The Northern Association preserved enormous and vital coal supplies, as well as being the buffer zone between England and the Scottish kingdom. The Southern Association again held vital supply routes from the south coast ports, and the vital foundries of Kent and Sussex, with its brother Western Association holding the strategic lines of communication stretching from Kent's ports, going along the channel, before turning through the Severn Valley. Each regional Association was important not only to its own boundaries, but was the web of supply to the central government and economic base in London. In this respect, each Association contributed to the war effort, but often in a disjointed and poorly coordinated way.[3]

When in desperation the Parliament's Lord General, Robert, Earl of Essex had requested reinforcements to back up his march into the West Country, the troops were unavailable, yet Parliament had found the time to debate the finances of the whole war.[4] Essex as Lord General was commander-in-chief, yet had little influence in the Associated counties, and even less direct command over commissions outside his own structure. This had been the pattern of administration since 1642, vast numbers of troops being raised, pay and supplies made available, but the transactions between Associations through committee, council and Parliament, had supplied a bureaucratic system whose sole objective was to balance the books.[5] But it was not only on pay and recruitment that the system was over complex: even more mundane everyday requirements could become major problems. An example of administrative chaos can be found in 1644, when Oliver Cromwell's regiment in the Eastern Association was quartered at Lincoln, and required both boots and horses from the Midland Association town of Northampton. For Cromwell could not buy his Ironsides boots upon a Parliamentary account, but had to pay in Eastern Association credits that would later be redeemable against money held by Association committees in London. Weeks were lost as first the need was assessed at military level, then the committee considered where to place the order, and then with manufacturers as their cottage industry work-force produced the boots. There can be little wonder at the numerous letters from civil war commanders, imploring committees to 'speedily send' clothing to their barefooted, ragged-britched soldiers. Needless to add, by the time the Eastern Association had transferred its credits through internal channels, the actual payments were long overdue.[6]

This logistical problem with its severe difficulty in administration, caused each regional Association to picture itself in the mould of a self-contained unit; indeed primarily autonomous within the Parliamentary system. It is no small wonder therefore, that such a system resulted in ever greater demands for military autonomy made by each of the generals.

Divisions And Diversions

Since 1643 the Earl of Essex and Sir William Balfour[7] had been jealous of the 'other army' commanded by Sir William Waller and Sir Arthur Hesilrige.[8] This situation had deteriorated during 1644 to a stage where Essex thought Waller had caused his humiliation at Lostwithiel, by failing to march westward to his assistance. The whole sorry business had worsened when the Earl of Manchester had appeared somewhat tardy in his response to orders to rescue Essex's forces by marching swiftly south with the Eastern Association. The effects of this far from satisfactory position were expanded and amplified by the fiasco now witnessed at Newbury. Yet the problem at Newbury was not wholly of tripartite tensions between jealous generals. There were differences within the three armies themselves. The civil war had not formed differences in religious sectarianism or toleration: these had undoubtedly been carefully honed before a single shot was fired or pike levelled. And to the religious distinctiveness had been grafted political aspirations, which varied between moderate presbyterian officers like Essex, Balfour, Manchester and Crawford, and the increasingly hawkish Independents like Cromwell and Hesilrige. The more junior officers too were divided within lines of dove or hawk, while some of the more radical young bloods had already taken on the guise of Jesuit.[9]

This situation was largely exemplified in the army of the Eastern Association by a quarrel between Sergeant-Major-General Lawrence Crawford and Oliver Cromwell. Crawford was professional in his duties, yet – being a Scot – his rigid presbyterianism brought him in direct opposition to the sectaries.[10] The problems of this religious intolerance flared between Crawford and Cromwell, when the Scot acting under the *Solemn League and Covenant* arrested Lieutenant Packer, for the Anabaptist religious doctrine he was espousing within the Association army. Although Cromwell himself was no Anabaptist, he was sufficiently tolerant of the more extreme sects to find Crawford's actions anathema to his own Independent conscience. In a classic reply to the charges laid against Packer, Cromwell maintained, quite curtly, that 'the State in choosing men to serve it, takes no notice of their opinions'.[11] Packer was released and Crawford held Cromwell increasingly responsible for the lack of presbyterian orthodoxy within the ranks of the Association. The differences between Cromwell and Manchester were for their part less upon religion than military expedience, with perhaps a degree or two of pure jealousy. Following the victory at Marston Moor, Cromwell had been hailed a national hero by the more radical presses, whereas Manchester received criticism for firstly missing the opening shots, then failing to pursue Prince Rupert's retreating forces.[12] Likewise Crawford had failed in an attempt to storm York, leaving him with considerable egg on his face, something rather unpalatable for any professional soldier, and more than ever for this rather proud Scot.

The main differences between the three Eastern Association commanders deteriorated even further during the summer, until the Committee of Both

Kingdoms eventually called all three to London in a bid for a convivial settlement. If a consolatory meeting was the desire of the Committee, this was not to be seen, for Cromwell pre-empted the talks by demanding the resignation of Lawrence Crawford.[13] The dramatic demand failed, yet was so public that as a result there formed a definite pro Cromwell camp within both the military and Parliament. Indeed although it can be argued that Cromwell held the political high ground before this time, it was the summer of 1644 that put him on a political platform as a military hawk and champion of the sects, and that began to form a rough mandate as to what Cromwellianism grew to mean. The whole sorry affair not only gave heart to the King's cause, but meant that Essex's army was sacrificed on an Eastern Association altar around Lostwithiel.

With these bitter tastes still high upon the tongue, the events at Newbury added further to Cromwell's displeasure. At a council of war following Newbury, Cromwell alleged that the Earl of Manchester had argued against pursuing the King, by stating that, 'if we beat the King ninety-nine times, yet he is King still, and so will his posterity be after him, but if the King beat us once we shall all be hanged, and our posterity be made slaves'.[14] A finer case for appeasement could hardly be made, yet this had been what some have believed truth since 1642. If Manchester believed that the King could not be brought to a settlement by military means, then why were they in the field at all?

The strained relationships within the Eastern Association brought to a head the fundamental need to reorganise the Parliamentary forces. Essex had already lost good officers to Manchester and Waller, and following the debacle of the summer and autumn, Essex's main Parliamentary army was re-equipped, yet its morale was in a very low condition. Essex and his Lieutenant General, Sir William Balfour, had supported the taking of the Covenant, and had strictly observed adherence to it in their forces, through which they had not suffered the same degree of religious disruption seen in Manchester's army. Conscientious differences had, however, caused some officers and men to re-enlist in Cromwell's more sectarian faction of the Eastern forces. One such was John Lilburne, a hero of Essex's 1642 army, but a man whose conscience had led him to seek purity of thought and deed under Manchester and Cromwell.[15] A less prominent, yet staunch, Parliamentary supporter, was William Allen. He was by birth a Warwickshire man, but before the civil war he had become a feltmaker in Southwark. It would seem that Allen first enlisted in Colonel Denzil Holles regiment in 1642 and fought at Edgehill and Brentford alongside John Lilburne. When Holles's men broke under extreme pressure in the fields around Brentford, Allen was captured, and was among those threatened with stigmatisation, or branding on the cheek, as a common traitor. Although the King had called for this punishment for those taken at Brentford, he later relented for political reasons.[16] Upon his release, Allen re-enlisted in Captain Beton's company of Philip Skippon's regiment, and was wounded at the first battle of Newbury and again at Henley, but following the imposition of the Covenant oath he left Essex's service.[17] This loss of troops by Essex to the Eastern Association, was matched by a similar exodus to Waller's forces, until by November 1644, the earl and his supporters began to grow fearful of his position.

Following the second battle of Newbury, the King resolved to relieve the siege at Basing House, where the Marquis of Winchester had held out for the royalist cause throughout 1644. However, the siege was raised by those besieging the garrison, and the two forces 'both retreated to their Winter-Quarters, the King's to *Oxford, Marlborough, Basing, Odiam, Newbery, Blewbery, and c.* The Parliaments at *Reading, Henly, Abbington, Farnham and c.*'[18]

With the final chance to engage the King's forces in decisive field combat prevented by Manchester, the stage was set for Cromwell and Waller to hasten to London and regain the limelight by accusation and political intrigue. Rumours of the real extent of the Newbury failure had already spread via Sir Arthur Hesilrige in the higher echelons, and no doubt were common gossip in the London taverns has the Trained Bands returned home. Immediately after the battle, Hesilrige had ridden to London, and appeared in the Commons, according to Denzil Holles, 'in beaten buff' to give his account of the events around Newbury. He had at this time exuded the gallantry of all the commanders, but shortly afterwards began the campaign against Manchester. Had the story of Manchester's reluctance to fight emulated solely from Hesilrige and Cromwell, it would be easier to discount, but added to their claims must be the fact that Sir William Waller was equal in his criticism. Waller was by religion a mild-mannered English presbyterian, in that group of politicians where the war was judged an unfortunate necessity, yet unlike Manchester he had seen service as a young man on the continent of Europe, and despite fighting in a 'war without an enemy' had ventured his life on numerous occasions since Edgehill. This extended period of loyal service had made Waller a national hero, and his position against the Earl of Manchester counted a great deal in the debate.

Returning to the Commons in November, Cromwell related his grievances to members eager to hear his words. Speaking as a man possessed by a sense of betrayal, Cromwell told of,

> his Lordships continued backwardness to all action, his averseness to engagement or what tends thereto, his neglecting of opportunities and declining to take or pursue advantages upon the enemy, and this [in many particulars] contrary to advice given him, contrary to commands received, and when there had been no impediment or other employment for his army.[19]

This reference from Cromwell's speech was not entirely in respect of Newbury, for he repeats in the words 'contrary to commands received', the earlier charge that Manchester had failed to relieve Essex at Lostwithiel. In this respect, the events following Newbury must not be viewed in isolation. Newbury was the match which ignited a fuse laid throughout that summer, not the cause nor the whole reason for such disquiet. Cromwell could through his ability to kindle loyalty, manoeuvre a political advantage for himself and his party. In this speech Cromwell made it quite clear that he doubted if Manchester would in future 'prosecute unto full victory' the war they were engaged upon. Yet unlike others of his generation, Cromwell never accused his political or military enemies of cowardice, this was not his way, to do so would reduce the charges to a personal level, whereas to question Manchester's willingness to fight the King, meant he could attack the Earl on a political stage.[20]

John Rushworth includes a draft synopsis of Cromwell's speech in his collection, in which he says,

> That the said Earl hath alwayes been very indisposed and backward to Engagements, and against the ending of the War by the Sword; and for such a Peace to which a Victory would be a disadvantage, and hath declared this by Principles express to that purpose, and a continued Series of Carriage and Actions answerable. And since the taking of *York* [as if the Parliament had now Advantage full enough] he hath declined whatever tended to further Advantage upon the Enemy, neglected and studiously shifted off opportunities to that purpose, [as if he thought the King too low, and the Parliament too high] especially at *Dennington Castle*. That he hath drawn the Army unto, and detained them in such a Posture, as to give the Enemy fresh Advantages, and this before his Conjunctions with the other Armies, by his own absolute Will, against or without his Council of War, against many Commands from the Committee of both Kingdoms, and with Contempt and vilifying those Commands. And since the Conjunction, sometimes against the Council of War, and sometimes perswading and deluding the Council to neglect one Opportunity with pretence of another, and that again of a third, and at last by perswading that it was not fit to fight at all.[21]

That the Earl felt unable to prosecute the war to the same extent as Cromwell is perhaps on reflection not too surprising, the whole concept of peerage being founded in kingship and subject, whereas the third estate of Commons is linked irredeemably with the people and citizenship. Although still in the future, for a peer to cut off a king's head would alter, fundamentally, the whole concept of the order of society, for an attack upon monarchy must be an attack upon peerage itself, because peers sit as Sir Arthur Hesilrige would later maintain 'by privilege alone'.[22] True, peers had killed kings in the past, but had followed their action by the imposition of a new dynasty, which already made the civil war different in its implications.

Manchester chose his own ground on which to resist Cromwell's charges. On 28 November he returned fire by a complaint upon his privilege, explaining the situation at Newbury and launching an attack upon Cromwell's radicalism. Their Lordships, never willing to suffer accusations against a member of their House, especially by a member of the Commons, called upon Manchester to record a statement to be presented on 2 December. This statement although long in content and equally long in wind, is worth examining in some detail, for it highlights many of the problems beset the Parliamentary military organisation. Manchester began his narrative thus:

> *My Lords,*
> The Trusts with which the Parliament of *England* have Honoured me, are of so great Concernment to the Publick, as I should be failing in the highest measure to your Lordships and myself as a Servant employed by you, if I should not be sensible of those Aspersions which Common Fame brings to my Ears, so as to endeavour to clear my self from that ignominious Brand of Unfaithfulness towards the Parliament, who have thought me worthy of their Favour and their Trust. Therefore I look upon this Command of your Lordships to give you an Account of my late Actions, not only as an addition to your former Favours, but as an advantage equivalent to my Life, for which I humbly offer your Lordships my acknowledgments as your Servant.
>
> *My Lords,*
> I shall not Plead my Abilities to serve you, I shall only justifie my Integrity in your Service, which if any shall contradict, if they be such as have either known me, or seen my Actions, when they shall question with their own Hearts, I doubt not but they will there find such results as will give them occasion to ask me Pardon for the Injury they have done me.
>
> *My Lords,*
> That which I hear gives the greatest dissatisfaction to the World in my particular, is the King's relieving Dennington-Castle, and the Armies not engaging with him; to this I shall make a Profession in general, that from the time I came to joyn with my Lord General's Army, I never did any thing without joynt Consent of those that were the best Experienced and chiefest Commanders in all the Armies; and herein I shall appeal to those who were sent down from the Committee of both Kingdoms, whether upon all Debates my Expressions were not these, I cannot pretend to have any Experience in this way, therefore what you shall resolve, I shall

observe; and I am confident that both they and all the Commanders of the Army will justifie my Practice made good my Professions.[23]

The Earl was calling for support not for his abilities, but for his lineage. As a peer, he called in his opening remarks for justification by service to the Lords. He was 'a servant employed by you' which suggests that as a peer he expected protection. Manchester then goes on to say 'upon all Debates my Expressions were not these, I cannot pretend to have any Experience in this way, therefore what you shall resolve, I shall observe', declaring before the Lords his lack of military ability. This point upon first examination appears quite startling, yet tells us of a fundamental weakness in seventeenth century armies, and confirms the point that lineage was stronger than ability in major decisions concerning senior command. Without saying more, the Earl confirms a class system in the military, he was a General not through ability, but through the right of peers to lead and command.[24] Turning to the battle of Newbury, the Earl continued,

> *My Lords,*
> At our first drawing up of our Armies towards *Newbery*, when the King lay there secured in his Quarters, it was resolved, that our Armies should be divided, that my Lord General's Foot and the City Brigade, with the most of all the Horse, should march to the West-side of *Newbery*, and that the Foot under my Command with some Horse, should remain on the East-side, and that as soon as I should by some Warning-pieces [artillery fire] see that they were engaged, that then I should make my Engagement for a Diversion. This Command was obeyed by me, and it pleased God through the Valour of my Lord General's Foot and some Horse, we had a very happy success of that Service; but where those Horses were that Lieutenant General *Cromwell* Commanded, I have as yet had no certain Account. After this [to omit our marching from *Newbery* towards *Abington*, and return thither again, all which was by the Advice and Consent of the Council of War] the King having gathered all his Forces together, draws them down towards *Wallingford*, and our constant Intelligence gave us, that he intended the Relief of *Dennington-Castle*; whereupon the *Thursday* my Intelligence being Confirmed, I sent unto Major General *Skippon* to Consult what was fittest to be done. We both resolved, that in regard all our Horse were quartered so far from us, it was necessary to call them to a Rendezvouz the next day, that so they might be nearer to us, and readier for any present Service. Hereupon Major General *Skippon* and my self writ to Sir *William Balfour*, that he would please to Command my Lord General's Horse to a Rendezvouz the next day, which he accordingly did. I sent likewise unto Lieutenant-General *Cromwell* to give the like Orders to my Horse, but he came unto me, and in a discontented manner expressed himself, asking me whether I intended to flea my Horse, for if I called them to a Rendezvouz, I might

have *their skins* but *no Service from them*. I told him my Opinion was, that it was absolutely necessary, for if it were not done, I doubted if we should have them present when we had most use of them; yet he persisting in his dislike of it, I told him he might do as he pleased. Upon the *Fryday* in the Evening we had certain Notice by a Lieutenant that came from the Enemy, that the King's whole Army was within five or six Miles. Hereupon we presently sent to Command all the Horse to be at Rendezvouz upon *Newbery-wash* by six of the Clock in the morning, intending to draw out to fight with the King. In order to which the ground was Viewed by the Chief Field Officers, but on *Saturday-morning* the King had gained his Passage to *Dennington-Castle* before any great Body of our Horse came up, so as it was resolved by all the Officers in chief, that it was fittest for us to stand upon our Defence, and to keep the Town of *Newbery*. About two of the Clock in the Afternoon the King Charged us with Horse and Foot near to the Works which we had made, but received a very happy repulse by our Foot; as yet there were only some of my Lord General's Horse, and some of Sir *William Waller's* came on that side the River that the Enemy was Lieutenant General *Cromwell* had not brought over any Horse, notwithstanding I had desired him that all of them might be drawn over on that side the River, where the present Service was. After some few hours that the Enemy had stood facing us, and that the Evening drew nigh, the Enemy through the Favour of the duskishness of the Evening made his Retreat, and about this time my Horse were coming into the Field, whereupon we all agreed that all the Horse should keep the Field that Night, and the Foot to make good their Posts as they had maintained them the day before, intending to draw out the next morning to attempt something upon the Enemy. In the Night we heard that the Enemy was marched away, whereupon Order was given by a general Consent, that the Horse should follow him by break of day; but in the morning certain Intelligence was brought us, that the Enemies whole Army was in a Body with in three miles of us, whereupon divers of us went to see whether it were true. And after we had rode about a mile to the top of a Hill, we saw the Enemies whole Army marching in an orderly Retreat. This gave Occasion to us all to consider what was fittest to be done, and most of the Commanders in the Army were called together, and there by general Consent it was agreed That it was not safe to engage against the King at the present. Many arguments were given. Sir *Arthur Haslerigg* used some Expressions to this effect, That we run a greater hazard than the King did, for if we beat him, his Army would not be ruin'd, but he being King still, and retreating to his Garrisons, he would recruit his Army, it now being the Winter season; but if he had the better of us, our whole Forces would be ruined, and the Kingdom in extream hazard, having no considerable reserve on this side *Newcastle*, so that the Enemy might without any opposition march up to the very Walls of *London*. And after some others had delivered their Opinions against Fighting, this Opinion of Sir *Arthur Haslerigg's* was seconded by me, and there was not one present that delivered his Opinion for fighting with the King at that time; and I conceive it was as far from our Intentions [as it was impertinent for the present purpose] to urge any of these Arguments as to the final result of the War; in the active and speedy Prosecution whereof, as I have often, so I shall still be willing upon all Occasions to hazard not only my self, but all that is dear unto me; but it was urged as not expedient to fight at that time, considering our present Posture, and by a general Consent it was thought fit to march back to *Newbury*.[25]

So far in Manchester's narrative he blames Cromwell for the backwardness of the Eastern Association horse, and puts into Sir Arthur Hesilrige's mouth a rewording of what the former accuses the Earl of saying, although it must be said for purely military reasons. The truth to this part of the whole affair may never be known, yet what is obvious from Manchester's words is that three armies under three Generals, working not under a chain of command, but by committee and consensus, was a disaster. The ability to carry out a military strategy was impossible under such circumstances, a point now evident from the narrative:

> When we had been some days at *Newbery*, we heard that the King intended to send a strong Party to relieve Basing, therefore the Council of War resolved that the best way to prevent any such Design, was to Order all the Horse of the Armies to keep guards there by turns, and though when the third Night came, that my Horse were to keep the guards, Lieutenant General *Cromwell* expressed an unwillingness to have any Horse to go to the Guards, yet I commanded that there should be no delay in it, and accordingly it was done. As for the several Motions of the Armies, and the drawing into those Quarters, where they now are, it was Ordered by the general Vote of the Council of War, not one differing. And I think I may with Confidence affirm, that there was such an Unanimity amongst us, as none acted any thing which was of Publick Concernment, apart from the Rest. As to that which may relate to me, I am fully perswaded that the Commanders in Chief will give me this Testimony, that I never concluded any thing without their Advice. And I must acknowledge that Lieutenant General *Cromwell* was sensible of a Contradiction in this particular, as when there was but an Information of such a Report came out at *London*, that I had acted without the Advice of the Council of War, he professed that he was a Villain and a Lyer that should affirm any such thing. That which I did without Consulting with the Commanders of the other Armies, was only such things as had a special respect to my own Forces, to keep them from mutinous Actions that they might be ready to commit, in regard of their great Necessities and Sufferings, and of several intimations which were given them that I was the only Caufe of keeping them there, and that Lieutenant General *Cromwell* was

willing and desirous to have them return to their Association. In this I confess I acted my own Power, to cause them to give Obedience to that which I had received Orders for from the Committee of both Kingdoms, though I shall ever shew as much readiness to serve for the security of the Association that have honoured me with their Favour, as any other shall do.

My Lords,
I hear further of a dissatisfaction which is of an older date, ever since my being at *Lincoln*, that when received Command from the Committee of both Kingdoms to march into the West, my backwardness was such, as I gave sharp Reproofs to those that mentioned it unto me. I cannot but wonder at such a Calumny, Lieutenant General *Cromwell* can Witness for me, that as soon as I received the Letters from the Committee, I consulted with him, and gave him Orders that twenty Troops of Horse should be got ready, and that he should go with them before me, and I would follow with the rest of the Horse and all the Foot with what possible speed I could. It is true that Lieutenant General *Cromwell* made some difficulties in regard of the Necessities that his Regiment of Horse were in, which I told him I would endeavour to supply at *Huntington*, and that I would send to London to make Provision of Boots and of other things which he wanted, and send them to meet him at *Reading*, and I made good my Promise, as many can bear me Witness. Certainly not only my Relation to my Lord General, to whom I owe both Honour and Service, but the Publick Interest, might justly Challenge from me a ready Obedience to this Service.

My Lords,
Some Discontents which then brake forth in my Army, was the Cause of retarding that Service. What those Discontents were, and the grounds of them, I dare not so far digress without your Command, as to offer them unto your Lordships.
I shall only now ask your Lordships Pardon for the Trouble I have given you, and shall beseech your Lordships to look upon me as one, who though I cannot serve you with Abilities equal to others, yet in my faithfulness to the Cause, in my Endeavours for the Happiness of the Parliament and Kingdom, and my Care of your Lordships Honours, shall give place to none.[26]

It is noticeable that throughout the narrative, Manchester pays little attention to Sir William Waller's forces, as if they were superfluous to the armies of Essex and the Eastern Association. Waller's army was certainly smaller than Manchester's force, and Essex's main field army, yet it was experienced and at times highly successful. This possibly unintended slight upon Waller and Hesilrige must have gone some way towards their vehemence to Manchester, for it is noticeable that Balfour and Cromwell receive orders from Manchester, but the Earl either dare not or could not extend such orders to the Southern and Western regiments.

At the same time the Earl of Essex, himself fearful of where Cromwell's views were taking their cause, tried to introduce Scottish influence into the fray by an attempt to persuade the Scottish Commissioners to bring charges against him as an incendiary. Essex, as we have seen, was a rigid supporter of the Covenant oath, Cromwell was not, and simply by not forcing his troopers to swear the oath could be accused of an 'attempt to divide the two kingdoms'. At a meeting of Presbyterian supporters called very late one evening by Essex sometime around 1 December, the Scottish Chancellor, urged that complaint be made of Cromwell's actions, and pressed the question that he might indeed be accused as an incendiary and brought to trial as one 'who kindleth coals of contention and raises differences in the State to the public damage.' It is interesting that John Rushworth leaves historians the following account, for it highlights the broad splits which were emerging in the Parliamentary cause, and further adds an insight to the rather tepid zeal of the Scots to meddle in English politics. Rushworth relates the Scot's words to the two Parliamentary 'learned members' Mr. Maynard and Mr. Whitlock whom Essex had also called to the meeting:

> I can assure you of the great opinion both my Brethren and my self have of your Worth and Abilities, else we should not have desired this meeting with you; and since it is his Excellency's pleasure, that I should acquaint you with the matter upon which your Counsel is desired, I shall obey his Commands, and briefly recite the Business to you.
> You ken vary weele that General Lieutenant *Cromwell* is no Friend of ours, and since the Advance of our Army into England, he hath used all underhand and cunning means to take off from our Honour and Merit towards this Kingdom; an evil Requital of all our Hazards and Services: But so it is, and we are nevertheless fully satisfied of the Affections and Gratitude of the gude People of this Nation in the General.
> It is thought requisite for us, and for the carrying on of the Cause of the Tway Kingdoms, that this Obstacle or *Remora* may be removed out of the way, whom we foresee will otherwise be no small Impediment to us, and the gude design we have under-taken. For he not only is no Friend to us, and to the Government of our Church, but he is also no well wisher to his Excellency, whom you and we all have cause to love and honour; and if he be permitted to go on his ways, it may I fear, endanger the whole Business; Therefore we are to advise of some Course to be taken for prevention of that mischief.
> You ken vary weele the Accord 'twixt the Two Kingdoms, and the Union of the Solemn League and Covenant, and if any be an *Incendiary* between the Two Nations, how he is to be proceeded against: Now the matter is,

> (wherein we desire your Opinions) what you take the meaning of this word *Incendiary* to be? And whether Lieutenant General *Cromwell* be not like an *Incendiary* as is meant thereby? and whilk way wud be best to tak to proceed against him, if he be proved to be like an *Incendiary*, and that will clepe his Wings from soaring to the prejudice of our Cause. Now you may ken, that by our Law in Scotland we clepe him an *Incendiary*, whay kindleth Coals of Contention, and raises Differences in the State to the publick damage, and he is *tanquant publicus Hostu Patria*: Whether your Law be the same, or not, you ken best who are Mickle Learned therein, and therefore with the favour of his Excellence, we desire your Judgments in these Points.[27]

Other Essex supporters were in high expectation that charges could be successfully brought against Cromwell. The moderates Denzil Holles and Sir Phillip Stapleton representing the Earl's opinion in the Commons, while Sir Henry Vane the younger and Sir Arthur Hesilrige backed their radical colleague. However, the two lawyers were of the opinion, that under English law, a pseudo-religious law like 'Incendiary' would be all but impossible to prove in a non church court, therefore to Essex's dismay, and Holles and Stapleton's visible anger, the lawyers concluded that such charges would fail in public court, and the Scots were 'not so forward to adventure upon it'.[28] The alternative was to bring a charge of treason against Cromwell, but this would have caused uproar within the cause itself, and it was concluded that if God could not commit him, the civil law was helpless to do more.

John Rushworth added an editorial note to this period, by saying:

> But Mr. *Whitlock* adds. *That there was cause to believe, that some present at this Debate were false Brethren, and informed* Cromwell *of all that passed, which might make him carry on his designs more actively for his own Advancement.* And indeed it may well be presumed he was not like to be behind hand in Artifices for removing of those that would have removed him.[29]

Although this counter-coup against Cromwell was a rather underground or clandestine affair, and despite the now possible danger to himself, he brought fresh charges against Manchester on 4 December, at a specially convened committee headed by Zouch Tate, formed by both Houses to find the acceptable truth, or knowing Parliamentary practises possibly to whitewash over the cracks appearing in their mutual cause. Sitting quietly through Manchester's defence, Cromwell rose to recite his charges, and presented them with considerable firmness of spirit.[30] The whole situation was falling into disarray, with charges and counter-charges, but had proven without any doubt that before a new season opened, the structure of the military required attention both in the field and administration.

NOTES

[1] Essex's health had suffered in the conditions in the field, and had not returned to his army. Although he would live for less than two years from this date, there is some justification in the view that this illness was only partly physical, and not a little political. Even less complimentary was the assertion by some royalist Newspapers that the Earl by this time preferred 'the fanny of his maid' to the rigours of the battlefield.

[2] *The Writings and Speeches of Oliver Cromwell*, ed. W. C. Abbott (4 vols., Cambridge, Mass. and London, 1937-47) vol. 1 [1599-1649], pp.302-1.

[3] The best account of an Association's administrative problems is C. Holmes, *The Eastern Association in the English Civil War* (Cambridge, 1974). See also the hundreds of directives to the Associations from Parliament and the 'Committee of Both Kingdoms', to be found in the Journal of the House of Lords, Journals of the House of Commons, and Calendar of State Papers Domestic series, hereafter, *L.J.*, *C.J.*, and *C. S. P. Dom.*, followed by volume and page number.

[4] Mark Kishlansky, *The Rise of the New Model Army* (Cambridge, 1979) pp.26-7.

[5] G. E. Aylmer, *The States servants: the civil service of the English Republic, 1649-60* (1973), gives a very comprehensive account of the progressive nature of the build up of State bureaucracy.

[6] It appears that the cottage industry shoemakers of Northampton were never paid for the 1644 order, although Cromwell, in accordance with military practice, paid for the horses via Lieutenant Russell. I am indebted to the Northampton Leathercraft Museum for this reference.

[7] Sir William Balfour was Lieutenant General of Horse to Essex's army from 1642. He was politically moderate, but did not have the Parliamentary influence of the Officer MP's like Cromwell, Stapleton, Waller or Hesilrige.

[8] Sir Arthur Hesilrige was Lieutenant General of Horse to Waller's army from December 1642, yet from 1644 held more political influence in Parliament, the Committee of Both Kingdoms and the Western Association, than did Waller.

[9] The word 'Jesuit', in this context, represented the preaching soldier just emerging particularly in the Eastern Association. They would grow in number during 1644-45.

[10] *The quarrel between the Earl of Manchester and Oliver Cromwell: and episode of the English Civil War*, eds. John Bruce and David Masson (Camden Society Publications, new series 12, 1875).

[11] Abbott (ed.), *Writings and speeches*, vol. 1, p.278.

[12] Bruce and Masson (eds.), *The quarrel*, p. xliii. It is interesting that Denzil Holles, a presbyterian colleague of Manchester, conveniently laid an identical accusation at Cromwell's part at Edgehill, in his post Restoration memoirs, *The memoirs of Denzil Lord*

Holles (1699). There does appear to have been a quite concerted effort to rewrite history at the Restoration which does not relate to the contemporary material available in letters and tracts.

[13] Robert Baillie, *The Letters and Journals of Robert Baillie 1637-1662*, ed. D. Laing (3 vols., Edinburgh 1841-2) vol. 2, pp.229-30.

[14] *C. S. P. Dom. 1644-5*, p.151.

[15] John Lilburne, *The legall fundamentall liberties of the people of England* (1649), extracts printed in *The Leveller Tracts, 1647-1653*, eds. William Haller and Geoffrey Davies (Second edition, Gloucester, Mass., 1964) pp.399-449, see esp. p.407. See chapter IV for a further discussion of this text.

[16] Sir Charles Thomas-Sanford, Sussex in the Great Civil War and Interregnum 1642-1660 (1910) p.86.

[17] Bodl. Tanner MS 58, f.84.

[18] John Rushworth, *Historical collections of private passages of state* (7 vols., 1659-1701) part III, vol. 2, p.732.

[19] Abbott (ed.), *Writings and Speeches*, vol. 1, pp.302-3.

[20] Ibid.

[21] Rushworth, *Historical collections*, part III, vol. 2, p.732.

[22] In a speech made in 1658, Hesilrige told how 'one could have Knights and Burgesses, but once have Lords and you have real privilege'.

[23] Rushworth, *Historical collections*, part III, vol. 2, pp.733.

[24] The idea that class played a large part in the English Civil War, is an highly contentious one, largely played down by Mark Kishlansky in *The Rise of the New Model Army*. I would suggest it was class by which the war now divided and progressed, yet not the traditional twentieth century struggle of left and right, but of trade against privilege.

[25] Rushworth, *Historical collections*, part III, vol. 2, pp.733-5.

[26] Ibid. pp.735-6.

[27] Ibid, part IV, vol. 1, p.2.

[28] Bulstrode Whitelocke, *Memorials of the English Affairs*, later edition ed. by Arthur Earl of Anglesey, (1732) pp.116-7.

[29] Rushworth, *Historical* collections, part IV, vol. 1, p.3.

[30] Abbott (ed.), *Writings and speeches*, vol. 1, p.313. Abbott concludes that 'military opinion was all but unanimous in supporting Cromwell's charges of Manchester's dilatory procedure; and had it not been, the result spoke for itself'.

CHAPTER II
TOWARDS SELF-DENIAL

With the quarrel between the earl of Manchester's supporters and the more hard-liners continuing unabated into December, it was becoming increasingly clear that it was not a question of right and wrong in a polarised argument, but a more fundamental problem of organisation and logistics. The methods of raising, organising, equipping and paying soldiers according to their army or Association status, was, for the main part, steeped in patronage and a perpetration of class government from previous centuries. In the balance of class order and governmental effect, the civil war was in its earliest years innovative in so much that it pitched Commons against their peers, but also relied upon the two estates of Commons and Peers working in unison to administer a war policy. This was true also for the King's party, yet if commoner and peer were united in their support for the monarchy, the structure of the court helped to preserve the balance which could not be so easily identified on the Parliament's side. It might be argued that the quarrel between Manchester and Oliver Cromwell was the result of the lordly second estate finding a socio-economic difference with the third estate of commoner, but by 1644 the whole structure of the war and even the Westminster Parliament itself was so intermixed that such a hypothesis is too simplified, although tempting in that simplicity.

Bearing this problem in mind, the chain of events set in motion by Zouch Tate MP on Monday 9 December 1644 are all the more surprising. After short debates concerning the 'poor captives of Algiers' and their likely fate in the galleys, and then a vote to empower the London Militia committee to search and detain deserters, the day's main business was called.[31] For a fortnight Tate and his committee for the reformation of the Army had considered the claims and counter claims between Manchester, Cromwell and Sir William Waller. Taking the floor, Tate, instead of reporting upon the findings of his committee, began by stating that 'pride and covetousness' were the principal problems within the Parliamentary cause, leading to jealousy and lack of coherence in its stratagem against the King.[32] It is not difficult to agree in hindsight with these remarks, the often quite petty remarks to be found within even official correspondence between army commanders and the Committee of Both Kingdoms are rarely shrouded in even the thinnest disguise, and the lack of response to the most direct of commands illustrates the frequently tempestuous nature of the relationships of senior officers.[33]

In what was probably a bid to plaster thinly over the deep cracks of division between the factions in the military, Tate reintroduced a proposal which had first appeared in 1643 to reform the military structure, and by so doing introduce a new army into the arena.³⁴ The army moreover, should not only be remodelled physically, but also receive a cleansing from the conditions of past ambition, by the barring of members of Parliament from military service, Tate stating,

> ...that during the time of this war, no member of either House shall have or execute any office or command, military or civil, granted or conferred by both or either of the Houses of Parliament.³⁵

For Zouch Tate to offer such a diagnosis of affairs before the Commons must have indicated to all assembled that the hawks or war party were prepared to offer a stand against the moderate prosecution of the conflict reflected in Manchester's actions at Newbury. One can imagine the hushed, almost shocked, silence that greeted Tate's statement, a silence broken by the East Anglian tones of Oliver Cromwell, who said,

> That it was now a time to speak, or forever hold the tongue. The important occasion being no less than to save a Nation out of a Bleeding, nay, almost dying condition, which the long

continuance of the War had already brought it into; so that without a more speedy vigorous and effectual prosecution of the War, calling off all lingering proceedings like Soldiers of Fortune beyond Sea, to spin out a War, we shall make the Kingdom weary of us, and hate the Name of a Parliament. For what do the Enemy say? Nay, what do many say that were friends at the beginning of the Parliament, even this, That the Members of both Houses have got great Places and Commands, and the Sword into their hands, and what by Interest in Parliament, and what by power in the Army, will perpetually continue themselves in Grandeur and not permit the war to speedily end, lest their own power should determine with it. This I speak here to our own Faces, is but what others do utter abroad behind our Backs. I am far from reflecting on any, I know the worth of those Commanders, Members of both Houses who are yet in power; but if I may speak my Conscience without reflection upon any, I do conceive if the Army be not put into another Method, and the War more vigorously prosecuted, the people can bear the War no longer, and will enforce you to a dishonourable Peace. But this I would recommend to your Prudence not to insist upon any Complaint or oversight of any Commander in Chief upon any occasion whatsoever; for as I must acknowledge my self Guilty of Over-sights, so I know they can rarely be avoided in Military Affairs; therefore waving a strict inquiry into the Causes of these things, let us apply ourselves to the Remedy which is most necessary; And I hope we have such true English hearts, and zealous Affection the general weal of our Mother country, as no members of either House will scruple to deny themselves and their own private interests for the public good...[36]

There can be little doubt that these were far from spontaneous words from Cromwell, for although he was noted for the ability to transform the moment, the subject of this matter was too important for it to have slipped innocently from the tongue.[37] Yet if spontaneity was not evident, a willingness to end the careers of Manchester and the old guard in return for patching over the division between himself and the peace group can be seen. That none would prosper from an inquiry into the events at Newbury, becomes even more clear at this time, and Cromwell was openly suggesting that if the barrel load of Christmas apples was examined thoroughly enough, more than a few rotten fruits would be found. Cromwell speaks of 'over-sights', and then delivers his strongest lines by appealing to the members' patriotism. The use of 'Mother England', 'English zeal' and 'self-denial' to the public interest are the words of Cromwell's trade, which was to transport the moment by appeals to man's better side, and in no small part, his natural vanity.

This question of purifying the army from outside considerations was in itself a constructive step. Obviously a war that became subject to profiteering was an abuse of civil power by political and military means. It would soon become apparent to the City merchants and tradesmen who were to no small degree subsidising the war, that Parliament was far from being their creature, and in a country where perpetual Parliaments were hardly the status quo, a return of the King might not seem so bad after all. That Cromwell should desire the purity of Parliament, speaks well of the intent to rid the seat of power of corrupting influences. It was said that Speaker Lenthall openly accepted bribes, and Sir Arthur Hesilrige's personal secretary would only arrange a meeting or accept a petition upon receipt of ten shillings.[38] Denzil Holles many years later accused Hesilrige of keeping his estate better stocked and protected during the war than ever it had been in peace time, although in fairness to Sir Arthur, his accuser had by the time of the accusation awaited his death and had himself accepted a post restoration peerage.[39] In the matter of Self Denial (as Cromwell's proposal would now be called) Hesilrige, a serving 'Lieutenant-General of Horse', sided with Cromwell and Tate, whereas Holles, by now a non-combatant member was more cautious.

It is probably true however, that neither Cromwell nor Tate were primarily responsible for the principles set forth in Self-Denial. In January 1644 the Parliamentary member for Hindon, the lawyer Robert Reynolds, suggested that most members of the Committee of Both Kingdoms held military positions in the Army or Associations, and while this situation remained they could continue the war for their personal advantage.[40] It is interesting that the younger Vane voted against Reynolds' early attempt to purify the war effort, but was a principal mover of Tate's almost identical Ordinance within that same year.[41] Following the words from Cromwell, a further – unidentified – member rounded off the proposal by saying:

> ...Tis apparent that the forces being under several great Commanders, want of good Correspondency among the Chieftains, has oftentimes hindered the public service...[42]

With this air of wonder floating amid the members, Tate formally moved his resolution, which was seconded by Sir Henry Vane Jnr, and was presented to an eight-man committee for drafting into an ordinance.[43]

The City, to which the war had become a financial burden, eagerly grasped Self-Denial as a way to reduce their taxation, bring the war to an ordered conclusion, and return the commanders back under civil control. D'Ewes writing from notes made earlier, recalls that 'some citizens of London' thanked the House for their action in passing the vote to bring forth the draft ordinance.[44]

The whole nature of self-denial was the spirit of the moment; there was something in its conception which caught the imagination of quite diverse political positions. Self-denial contained within it a purity of concept that could be grasped by all shades of the puritan ethic. It was to offer the panacea, which by its acceptance, all errors and jealousies of 1644 could be salved away. It was in its simplest form an example of political fasting, for by denying themselves of military service, the politician could be seen to be giving physical penance for his spiritual sins. That many saw the setbacks suffered in 1644 as a judgement of God was a common diagnosis of the nations ills, therefore to offer themselves as sacrifice was an answer to wipe away these sins, the act of self-denial giving to the politician of 1644-45 the perfect question to Cain's eternal answer 'I am not my brothers keeper.'

The reintroduction of Self-Denial was probably due in no small measure to the Commons own folly on 14 November in awarding John Lisle M.P the Mastership of St.Cross Hospital at Winchester.[45] This position offered with it a fee worth an estimated £800 a year, and the furore that erupted in the House was totally justified.[46] In response a committee was set up to,

> inquire into the value and nature of all the offices and places, and other advantages that have been bestowed by the Parliament, or by virtue or colour of any authority of Parliament; what is received by the persons on whom such offices, places, or other advantages are bestowed; and to consider what allowance is to be made out of them, to the persons that execute them; and what to the use of the publick: And are to begin first with the offices, places, or other advantages bestowed upon the members of either House.[47]

It is significant that ten months after first raising the notion of Parliamentary corruption in awarding official positions and the persecution of the war, the head of this committee was Robert Reynolds.[48] The original question set by Reynolds in January appertaining to the military eagerness in continuing the war, was at this same time reintroduced by a Petition by a group of Kentish knights, which states in the strongest terms:

> To the intent that this obligation of assistance [particularly pecuniary] so just and necessary yet so suitable to the soldiers' present interest of making a trade of war, may not prove an occasion of lengthening out our miseries, we shall humbly crave that some honourable and beneficial reward may be settled to the commanders and common soldiers, to be received by them out of the estates of delinquents at the end of the warre, as may quicken them to a noble desire of the speedy enjoyment thereof. And in the meantime such competent allowance only to be made to all commanders...as may reasonably defray the charge of their employment.[49]

The committee for Kent had long suffered problems administering its troops co-opted to Sir William Waller's army by the South-eastern Association. Even in November 1644 the Kent Horse led by Sir Michael Livesey were semi-mutinous and their commander under suspicion of cowardice. Kent was geographically in a unique position in relationship to the war, protected as it was by the Thames and the sea. To attack the free fields of Kent a royalist army had to sidestep London and enter by the east Sussex side door, cutting itself off from reinforcements. For this reason the taxpayers of Kent had found themselves supplying manpower to the Parliamentary forces, and money for the payment of the same, yet apart from singularly local uprisings, had not suffered major military disturbance like its counterparts in the western counties. But perhaps it was not such local problems which brought the lack of military success and the soldiers' reluctance to end the war in question, for it was certainly a subject taboo in all but the most outspoken journals. Only the *True Informer*[50] and *The Scottish Dove* dared to summarize the Kent petition.[51] Yet in the following weeks the licensed press did show some evidence against blatant profiteering. In reality the members of both Houses could hardly deny that office and financial reward were brothers under the same skin, before his death John Pym had been Master of the Ordnance which had paid off at least some of the debts his colonization investments had incurred over the years, and Vane's friendships in the city through his office as Treasurer to the Navy, had increased his standing in society which could only have proven beneficial to his finances.

Despite the danger in failing to acknowledge the discourse within its supporters minds, or perhaps because of that danger if the reality was openly discussed, the Kentish petition was dutifully ignored and covered more comprehensively by self-denial, which in the five day gestation period from conception on the 9th to its birth on the 14th took upon it a truth possibly missing from its paternity.

The Weekly Intelligencer found the equality in the justice of self-denial to be worthy of its support.[52] *The Scottish Dove* also found that 'by this vote the Parliament do clearly free themselves of all imputations of covetousness.'[53]

A puzzled and undoubtedly troubled Robert Baillie, looking for righteousness in their actions and searching for the panacea which we have found to be required if God was to once more show his grace upon their cause, wrote almost a month later:

> They have taken all office from all members of both Houses. This was done on a sudden, in one session, with great unanimity; is still more and more admired by some as a most wise necessary and heroic action; by others as the most rash, hazardous and unjust action as ever Parliament did. Much may be said on both hands but as yet it seems a dream and the bottom of it is not understood.[54]

That Baillie recognized the fairness in self-denial owes much to the infighting he had witnessed in his native Scotland, leading to the inertia of the Scottish system of government where family loyalty and self interest often replaced common sense and the necessity to pass vital decisions.[55]

Clarendon however, although biased against finding any honesty or truth in the Parliament's deeds, was rightly sceptical of their sincerity, and records that to sell the idea of self-denial the spiritual weight of the preacher's holy rhetoric was enlisted by the war party:

> They knew not how to propose the great alterations they intended to the Parliament; and of all men, the Scots' commissioners were not to be trusted. In the end, they resolved to pursue the method in which they had been hitherto so successful, and to prepare and ripen things in the Church, that they might afterwards in due time grow to maturity in the Parliament. They agreed therefore in the House (and in these combinations they were always unanimous,) that they would have a solemn fast-day, in which they would seek God, which was the new phrase they brought from Scotland with their Covenant, and desire his assistance, to lead them out of the perplexities they were in; and they did as readily agree in the nomination of the preachers who were to perform that exercise, and who were more trusted in the deepest designs than most of those who named them were; for there was now a schism among their clergy as well as the laity, and the Independents were the bolder and more political men.[56]

Clarendon was writing from earlier notes, and his own copious collection of letters and tracts, years after the events during his exile, but this period sees him introduce the terms Independent and Presbyterian to identify political groups. From this it would appear that the royalists, at this time, did recognise a polarization of political and religious stances into parties; however loosely that term applied, and distrusted almost all.

On 11 December the report stage of the Ordinance reached the Commons ready for its reading on the 14th, at the same time a Fast being ordered 'for imploring a Blessing on the intended New Model of the Army', and this second action was accepted and passed by the Lords the next day. At the sitting on the 14th, the Commons took the unusual step by turning themselves into a Grand Committee, thus freeing themselves from the restraints set by Parliamentary procedure, a committee being at liberty to openly discuss every aspect of the task in question without referral to other authority. The promoters of self-denial maintained,

1. That upon passing this Ordinance the proceedings of their Armies would be more quick both Determination and Action, when Commanders upon any discontent arising, shall have the less power to Sway and Bandy one against another.
2. If there be at present Differences between Commanders that are Parliament men, and perhaps of several Houses, by this Ordinance equal Justice is done, they are both recalled from Command, and by consequence from further dispute or difference, thereby preventing Divisions in the Army, administring Advantages to the Enemy.
3. The Commanders will be the less able to make parties to secure themselves when they have no Interest in the Houses, and so become more easily removable or punishable, for their Neglects and Offences committed in the Army.
4. His Majesty by his late Message having acknowledged this to be the Parliament of England, both Houses had need to be as full as they can; and by this Vote the Members taken off from other imploys will be better able to attend the Publick Affairs in Parliament, to which they are called by Writ; and so the frequent Objection of the Paucity of Members at the passing of Votes will be prevented.
5. That by the New Model designed, the former Weights that obstructed being taken off, the progress of the Army upon new Wheels will be more swift, and a speedy period put to the War, which is the general desire of the Nation.
6. Hereby the Objection, That the Members seek their own Profit, Honour, and Power, will be prevented; when the World shall see them so ready to exclude themselves from all Commands and Offices.[57]

Of these reasons only section four offers a new reason to restrict the service of Parliament men in military or other office, this being, the lack of members in both Houses causing stress upon the quorum. Since the deaths of such men as Pym, Hampden and Brooke, the absence in the field of such others as Waller, Hesilrige, Stapleton and Cromwell from the Commons, and Lords Grey, Fairfax and Manchester from that House, had placed a strain upon the legality of Acts and Ordinances passed by this rump of the Long Parliament.[58] The reformers however were not totally united in their appraisal for the method of filling the empty benches, for a speech attributed to Bulstrode Whitelock by John Rushworth, delivers another interpretation on how best to reform the war:

> Mr. Speaker,
> I am one of that Number of your Servants, who have no Office or Imployment but such as you are about to except out of this Ordinance, nor have any ambition for any; and therefore may the more freely and indifferently, yet with all Submission, humbly offer my Reasons against it, as that which I apprehend may prove prejudicial to your Service.

It hath been Objected that your House, and the House of Lords, is thin and empty, and you the less esteemed having so few Members here, many of them being imployed in Offices that they cannot attend the Houses; but that by this Ordinance they will be at leisure and liberty to attend the Service of the Parliament here, and the Houses be much fuller than now they are.

I confess, Sir, this is fit to be remedied; but I apprehend you have a fitter way, than by this Ordinance to do it; that is, by issuing out new Writs for Electing new Members in the Places of those who are dead or expelled, and this will satisfie the Objection, and engage divers of Interest and Quality, the more immediately in your Service; whereas this Ordinance will discontent many, and the Houses will be little the fuller by the passing of it.

Another Objection is, that if this Ordinance doth not pass, the Treaty of Peace will not so well proceed, but the particular Interests of Members of Parliament may retard the same, but will all be taken away by this Ordinance.

I am to seek how this can be materially objected, when I suppose, whether this Ordinance pass or not, yet you intend Members of Parliament to be your Commissioners for that Treaty, and in case some of them be Officers, they will the better understand your Businesses, on which the Treaty will be grounded.

Another Objection is, That unless this Ordinance pass, the Great Work intended of New Modelling your Armies will not be so well carried on, for that by putting all out there will remain no exception.

I should rather have argued, that by putting out all Members out of their Imployment, the exception and discontent would be the more General, and by leaving them still in their Imployment, there would be the less Competition and Sollicitation for new Officers in their Rooms.

Another Objection or Argument is that the Members of Parliament, who are Officers, being of equal Power in Parliament will not be so obedient to your Commands as others who have smaller Interests, and would not so much dispute one with another.

Surely, Sir, those whose Interest is the same with yours, have the more reason to obey your Commands than others, and have more to hazard by disobedience, than others can have, and in your Commands all your Members are involved, and it is strange if they should be backward to obey their own Orders.

Nor will the contests be so frequent and high between them and other Officers, as it will be between those who will be of a more equal Condition.

But Mr. Speaker, as you consider the Inconveniences, if this Ordinance do not pass, so you will be pleased to consider the Inconveniences if they do pass.

You will lay aside, as brave Men, and who have served you with as much Courage, Wisdom, Faithfulness and Success as ever Men served their Country.

Our Noble Generals, the Earls of Denbigh, Warwick, Manchester, the Lords Roberts, Willoughby and other Lords in your Armies, besides those in Civil Offices not excepted; and of your Members the Lord Grey, Lord Fairfax, Sir William Waller, Lieutenant-General Cromwell, Mr. Hollis, Sir Philip Stapleton, Sir William Brereton, Sir John Meyrick, and many others must be layed aside if you pass this Ordinance.

And I am to seek, and doubt so will they be, to whom you shall refer the New Modelling of your Armies, where to find Officers that shall excel, if equal these.

If your Judgements are that for the Publick Service it will be expedient to remove any of them from their Commands, let the same be plainly made known to them from you.

Let them have what they deserve, your Thanks for their former good Services, and they will not be offended, that you having no more Work for them, do lay them aside with Honour.

But to do Business of this nature [as hath been well said] by a sidewind, is in my humble opinion, not so becoming your Honour and Wisdom, as Plainness and Gravity, which are Ornaments to your Actions.

I shall conclude with the Examples of the Grecians and Romans, amongst whom, Sir, you know that the greatest Offices of War and Peace were conferred upon their Senators, and their Reasons were because they having greater Interests than others, were the more capable to do them the greatest Service. And having the same Interest with the Senate, and present at their Debates, they understood their Business the better, and were less apt to break that Trust, which so nearly concerned their private Interests which was involved with the Publick, and the better they understood their Business the better Service might be expected from them.

Sir, I humbly submit the Application to your Judgement. Your Ancestors did the same, they thought the Members of Parliament fittest to be imployed in the greatest Offices. I hope you will be of the same Judgement, and not at this time to pass this Ordinance, and thereby to discourage your faithful Servants.[59]

The conservative nature of Whitelock is exemplified in his desire to preserve the natural order of the three estates. He could foresee the problems an Army ordered by 'the middle sort of person' could bring, inadvertently seeing in the frame of the New Modelling the lack of discipline from above that would lead to the crisis of 1647. If the appointment of the Earl of Essex in 1642 had given Parliament itself the freedom to wage war against its enemies, then Whitelock could envisage that an Army free from the restraints of its natural hierarchy, could, given the circumstances, be uncontrollable by those for whom it was assembled. In reality, Whitelock was unable on this day to accept or differentiate between the State as a nation, and the nation of an ordered or structured State. The question of for whom was the war being fought was a relatively new one, and not to be answered by self-denial, but in its setting was to ask such constitutional riddles that have remained unsolved to modern times.

The debates continued only to be broken by the Fast-day which fell on 18 December and was held in Lincolns Inn Chapel, to a closed congregation with 'no Strangers, nor so much as their own Attendants to be admitted.'[60] All this occurring just over a week after Tate's first motion, yet this was long enough for the newspapers to inform their eager readers and for the preachers to prepare their congregations for the noble deeds and pseudo-holistic miracles about to emerge from the mouths of politicians. Clarendon believed none of it, the motives of those now pressing for self-denial were according to the critical earl, highly dubious at best, and dishonest at their final judgement. He continued to write:

> When the fast-day came, which was observed for eight or ten hours together in the churches, the preachers prayed the Parliament might be inspired with those thoughts as might contribute to their honour and reputation, and that they might preserve that opinion the nation had of their honesty and integrity, and be without any selfish ends, or seeking their own benefit and advantage. And after this preparation by their prayers, the preachers, let their texts be what they would, told them very plainly, that it was no wonder that there was such division amongst them in their counsels when there was no union in their hearts; that the Parliament lay under many reproaches, not only amongst their enemies but with their best friends, who were the more out of countenance because they found that the aspersions and imputations which their enemies had laid upon them were so well grounded that they could not wipe them off; that there was so great pride, as great ambition, as many private ends, and as little zeal and affection for the public, as they had imputed to the Court; that whilst they pretended, at the public cost and out of the purses of the poor people, to make a general reformation, they took great care to grow great and rich themselves; and that both the city and kingdom took notice, with great anxiety of mind, that all the offices of the army, and all the profitable offices of the kingdom, were in the hands of the members of the two Houses of Parliament, who, whilst the nation grew poor, as it must needs do under such insupportable taxes, grew very rich, and would in a short time get all the money of the kingdom into their hands; and that it could not reasonably be expected that such men, who got so much and enriched themselves to that degree by the continuance of the war, would heartily pursue those ways which would put an end to it, the end whereof must put an end to their exorbitant profit. And when they

had exaggerated these reproaches as pathetically as they could, and the sense the people generally had of the corruption of it, even to a despair of ever seeing an end of the calamities they sustained, or having any prospect of that reformation in Church and State which they had so often and so solemnly promised to effect, they fell again to their prayers that God would take his own work into his hand, and if the instruments he had already employed were not worthy to bring so glorious a design to a conclusion, that he would inspire others more fit, who might perfect what was begun, and bring the troubles of the nation to a godly period.[61]

One can but imagine the fervour of the preaching, and the electric atmosphere whipped up in the assemblies. Although a fast of ten hours would hardly show itself upon the ample girths of many, this self righteous starvation in the service of God, would convince many an empty taxpayer that if belts were to be tightened, a good place to start might be in the Parliament and military hierarchy, with the army metaphorically marching on its own stomach rather than the taxpayers.[62]

In the debate held on 19 December, the day after the great Fast and prayer meeting, Sir Henry Vane Junior entered into a long speech of recommendation favouring self-denial, ending his speech with an offer to end his treasurership of the Navy.[63] It is however interesting to note that Vane did not lose his position owing to a clause which exempted from self-denial those who had been dismissed from office by the King, but subsequently reappointed by Parliament.[64] It is almost without question that Vane offered himself as the human and monetary sacrifice to further the Ordinance's passage from the Commons to the less certain waters of the Lords. From its introduction in the Commons on the 14th, the Ordinance had taken five days, including one Sabbath day and the fast day, which means self-denial was rushed through the Commons in a matter of two and a half to three days maximum.[65] With this taken into consideration, it is simpler to accept how the Independents like Vane and Cromwell could orchestrate their positions to carry the Ordinance quickly into its report stage in the upper House. Speed was the political tool which the war party was to use time and again to cut short the debate, a tool which had carried through the first Committee of Both Kingdoms earlier that year, and indeed had been the defence of Parliament in the days leading up to the attempted arrest of the 'five members' in 1642.[66] But with such fundamental changes on the self-denial agenda, could the Lords bow to their duty and verify the Commons motions, or would the old order declare its own interest sacrosanct and split their cause asunder.

NOTES

[31] Mark Kishlansky, *The rise of the New Model Army* (Cambridge, 1979) p.28.

[32] Ibid. p.29.

[33] *C. S. P. Dom 1644*, and *1644-45*, for numerous examples of the lack of supportive action between Essex, Waller and the Eastern Association forces. It is noticeable that the small Northern Association under Ferdinando Lord Fairfax, was not stricken by these substructures within its understanding of the Parliamentary cause. The problem was not only to be found in the Parliamentary army however, for it had been a problem within Royalist ranks since 1642.

[34] Adair, John, *Roundhead general: a military biography of Sir William Waller* (1969) pp.13, 92-3. Both the Parliament and the old pre Covenant assembly, the Committee of Safety, had nurtured plans to place Waller over a highly organised new general purpose field army, but the jealousy of the earl of Essex, coupled with Waller's disaster at Roundway Down, had swayed the Parliament against such a move.

[35] Kishlansky *The rise of the New Model Army*, p.29.

[36] Oliver Cromwell's speech, 9 December 1645, quoted in John Rushworth, *Historical collections of private passages of state* (7 vols., 1659-1701), part IV, vol. 1, p.4.

[37] For a detailed analysis on Cromwell's part in Self Denial, see A.N.B. Cotton, 'Cromwell and the Self-Denying Ordinance', *History* 62 (1977) pp.211-31.

[38] H. N. Brailsford, *The Levellers and the English Revolution*, ed. Christopher Hill, (Spokesman University Paperbacks, no. 14, Nottingham, 1976) p.2, [originally published by Cresset Press, 1961].

[39] Denzil Holles, *Memoirs of Denzil Lord Holles* (1699) pp.139-40.

[40] 'The parliamentary diary of Lawrence Whitacre', BL Add. MS 31,116, f.113v.

[41] Ibid. Lawrence Whitacre records that Vane opposed Reynolds at this time, but on other occasions the war party to which Vane adhered himself used Reynolds to introduce proposals advantageous to their stance.

[42] Anonymous speech, 9 December, 1645, quoted in Rushworth, *Historical collections*, part IV, vol. 1, p.4. The tone of an earlier part of this speech is somewhat poetic, talking of 'A Summers Victory has proved but a Winters Story; the Game however shut up with Autumn, was to be new played again the next Spring; as if the Blood that has been shed, were only to manure the Field of War for a more plentiful Crop of Contention,' ibid. In many respects the poet soldiers of 1914-1918 or Civil War Spain in the 1930s might have identified with the jaundiced view of this politician.

[43] Ibid.

[44] 'A collection of papers, containing a continuation of Sir Simonds D'Ewes's Journal of the House of Common from 13 Feb 1643 to 3 Nov 1645', BL Harleian MS 166, f. 151. D'Ewes wrongly dates this 28 October but it should almost certainly dated after 9 December 1644.

[45] Whitacre's diary, BL Additional MS 31, 116, f.174.

[46] BL Thomason Tracts E.465[19], *Mercurius Pragmaticus*, No. 27 (26 September-3 October 1648) in retrospect gives the assessment thus, Aulicus can however, be highly suspect in its figures.

[47] *C.J.*, vol. iii, p.695.

[48] Ibid.

[49] BL Thomason Tracts E.19[11], *The Kentish petition presented about the beginning of November 1644*, quoted in Rowe, Violet A., *Sir Henry Vane the Younger: A study in political and administrative history*, (University of London Historical Studies, no. 28, 1970) p.56.

[50] BL Thomason Tracts E.17[9], *The True Informer*, No. 54 (9-16 November 1644).

[51] BL Thomason Tracts *The Scottish Dove*, No. 57 (15 -22 November 1644) E.18[7]. George Thomason could only obtain a hand-written copy of the Kent petition, (see n. 19 above) and added to it a note that 'Mr. Rushworth durst not license it to print.'

[52] BL Thomason Tracts E.21[11], *The Weekly Intelligencer*, no. 84 (3-10 December 1644), quoted in Kishlansky, *The rise of the New Model Army*, p.30.

[53] BL Thomason Tracts E.21[17], *The Scottish Dove*, no. 60 (6-13 December 1644), quoted in ibid.

[54] Ibid.

[55] For a detailed study of Scottish Parliamentary government, see David Stevenson, *Revolution and Counter-Revolution in Scotland*, D. Stevenson (1977).

[56] Clarendon, Edward Earl of, *The history of the rebellion and civil wars in England*, ed. W. D. Macray (6 vols., Oxford, 1888) book VIII, vol. 3, p.191.

[57] Rushworth, *Historical collections*, part. IV, vol. 1, p.5.

[58] Although the term 'Rump' is usually used at a later date, the reduction in members from 1640-1644 must suggest using this term as a paradox.

[59] Rushworth, *Historical collections*, part. IV, vol. 1, pp. 6-7.

[60] Ibid. p.5.

[61] Clarendon, *The history of the rebellion*, book VIII, vol. 3, p.192.

[62] There is undoubtedly a similarity in the fervour evoked in seventeenth century sectarian congregations, and the openly schismatic tendency of twentieth century fundamentalism. It is easy to assess the effect such rhetoric power can have upon political stability.

[63] Clarendon, *The history of the rebellion*, book VIII, vol. 3, p.194.

[64] 'The Self-denying Ordinance', 3 April 1645, printed in *The constitutional documents of the Puritan Revolution 1625-1660*, ed. S. R. Gardiner (Revised edition, Oxford, 1906) p.288.

[65] The Ordinance passed the Commons on 19 December but was not delivered to the Lords until the 21st.

[66] The modern method of using speed to increase the passage of Parliamentary business against a possible objection from an opposition faction is known as a 'guillotine motion' and it has the effect of cutting short the debate as in 1644.

CHAPTER III
A NEW MODEL FOR THE ARMY

Writing after the war, the presbyterian minister Richard Baxter expressed the position of the Earl of Essex at the time of the Self Denying Ordinance:

> And it was discovered that the Earl of *Essex's* Judgment (and the wisest Men about him) was never for the ending of the Wars by the Sword, but only to force a Pacificatory Treaty: He thought that if the King should Conquer, the Government of the Kingdom would be changed into Arbitrary, and the Subjects Propriety and Liberty lost: And he thought that if he himself should utterly conquer the King, the Parliament would be tempted to encroach upon the King's Prerogative, and the Priviledges of the Lords, and put too much Power in the Gentries and the Peoples hands, and that they would not know how to settle the State of the Kingdom, or the Church, without injuring others, and running into Extreames, and falling into Divisions among themselves. Therefore he was not for a Conquest of the King. But they saw the Delay gave the King advantage, and wearied out and ruined the Country, and therefore they now began to say, that at *Edgehill*, at *Newbury*, and at other times, he had never prosecuted any Victory, but stood still and seen the King's Army retreat, and never pursued them when it had been easie to have ended all the Wars.[67]

This danger of redistributing power as between English social classes was one with which the Earl was preoccupied. The gentry, perhaps typified by Cromwell and knights like Hesilrige and the younger Vane, were growing in national strength through trade and foreign investment. Although Cromwell himself had only limited involvement in the English colonial activities, the lower order of gentry were becoming important through their entrepreneurial investment in new areas of commerce. It is not difficult to identify the point therefore, that the Earl of Essex was deeply concerned with the position of sovereignty within the nation, and whether victory over the King would strengthen the gentry even further over the peerage.

The addition of the Earl of Northumberland's secretary, Robert Scawen, to head a committee for Army contracts, also seriously wounded the ability of Essex to control the direction that the new force would take. Northumberland himself was the natural opponent of Essex's baronial power within Parliaments internal structure, and siding with Saye could force issues through administrative committees, where the Lord General was less powerful. The new army was therefore a nucleus for true civil service power, where employed administrators could influence the war in a way previously enjoyed by the so called 'old blood' of the higher echelons of the aristocracy. War had been a mark of baronial power since the much-heralded conquest of 1066, and it is no surprise to find at this time a theme of the protective nature of the peerage in securing the liberty of the subject against monarchic excesses. The freedom of the Anglo-Saxon law, was, quite inexplicably, woven into folklore together with the Barons opposition to King John. The Earl of Essex therefore was in 1642 the champion of baronial power against King Charles, and at the same time the upholder of constitutional liberty of the subject.[68] In the eyes of Essex therefore his power was the power of inheritance, the right of command was his through generations of the baronial inheritors' protection of the lower classes. The new army was a direct threat to this power, and in Baxter's view was seen by Essex as a challenge to the prerogative of King's and to the privilege of the peerage. As we have already seen, Essex had tried in December to indict Cromwell as an 'Incendiary', which Whitlock and Maynard had found to be unworkable in English law. According to Rushworth the lawyers had also related of Cromwell:

> That for their own parts they take Lieutenant General *Cromwell* to be a Gentleman of quick and subtle parts, and one who hath (especially of late) gained no small Interest in the House of Commons; nor is he wanting of Friends in the House of Peers, nor of Abilities in himself to manage his own part or defence to the best Advantage.[69]

Similarly the question of religion also came high on the list of important factors in the new army. Baxter again relates the situation between Essex and the Independent's view:

> Sir *H. Vane* by this time had increased Sectaries in the House, having drawn some Members to his Opinion; and *Cromwell*, who was the Earl of *Manchester's* Lieutenant General, had gathered to him as many of the Religious Party, especially of the Sectaries as he could get; and kept a Correspondency with *Vane's* Party in the House, as if it were only to strengthen the Religious Party: And *Manchester's* Army, especially *Cromwell's* Party, had won a victory near *Horncastle* in *Lincolnshire*, and had done the main Service of that day at the great fight at *York*; and every where the Religious Party that were deepliest apprehensive of the Concernment of the War, had far better Success than the other sort of Common Soldiers.[70]

Richard Baxter was no supporter of the sectaries, so it is therefore interesting that he freely acknowledges that 'the Religious Party' was more successful in prosecuting the war. In a further reference, Baxter emphasised this point by stating the nature of the men employed by Cromwell:

> At his first entrance into the wars, being but a captain of horse, he had a special care to get religious men into his troop: These men were of greater understanding than common soldiers, and therefore were more apprehensive of the importance and consequence of the war; and making not money, but that which they took for the publick felicity, to be their end, they were the more engaged to be valiant; for he that maketh money his end, doth esteem his life above his pay, and therefore is like enough to save it by flight when danger comes, if possibly he can: but he that maketh the felicity of Church and State his end, esteemeth it above his life, and therefore will the sooner lay down his life for it. And men of parts and understanding know how to manage their business, and know that flying is the surest way to death, and that standing to it is the likeliest way to escape; there being many usually that fall in flight, for one that falls in valient fight. These things it's probable Cromwell understood; and that none would be such engaged valiant men as the religious: But yet I conjecture, that at his first choosing such men into his troop, it was the very esteem and love of religious men that principally moved him; and the avoiding of those disorders, mutinies, plunderings, and grievances of the country, which deboist men in armies are commonly guilty of: By this means he indeed sped better than he expected. Aires, Desborough, Berry, Evanson, and the rest of that troop, did prove so valiant, that as far as I could learn, they never once ran away before an enemy.[71]

It must be said that Cromwell had indeed gathered the more sectarian elements of the armies to him, ranging from the Anabaptist like Lieutenant William Packer who was now a Captain in his own regiment of horse, to the radicals like John Lilburne who had commanded Manchester's dragoons. Whether it is totally safe, or indeed wise, to term Cromwell's troopers zealots, jesuits or fundamentalists, is open to question, but he appears to have found men with a like disposition to his own superior to those seeking personal glory and self interest. By placing religious men in his regiment, he had formed a strength which can be likened to the loyalty shown by the more devout royalists to the King. Bulstrode Whitelock confirmed in his memoirs, that Cromwell chose 'most of them freeholders, and freeholders sons, and who upon matter of conscience engaged in this quarrel under Cromwell'.[72] For some months rumours had circulated that the Independents would use the sectaries in the army to bolster their position, although Cromwell had denied it. A further report on Cromwell's 25 November speech against Manchester highlights his avowed loyalty to public order and the law. Henry Walker, publisher of *Perfect Occurrences*, recalled:

> This day the House of Commons spent much time about the business of the Armies not fighting at Donnington, and General Cromwell spake very worthy (as also Sir William Waller), General Crumwell in his speech certified as followeth.
> The heads of Generall Crumwell's Speech.
> Whereas it is reported that the Independents would not fight, that it was a scandall put upon them, and that none can say, but than every man of them very desirous to fight, and that whereas some fear that the Independents would be troublesome, when the Presbytery was settled (that he beleeved) that they would live under their laws quietly, and be obedient either actively or passively. And for those who were charged to be the cause why they did not fight, it being referred to examination, it is not thought fit to name any person of honour until the businesse be further examined.[73]

A rival newspaper, *The Parliament Scout*, failed totally to record these views held by Cromwell, yet in the words of Henry Walker can be seen the pseudo-religious politics exemplified years later in Baxter's prose, and the writings of another theologian Thomas Edwards.[74] The '*Scout*' did however, in an editorial on 5 December, try to paper over the cracks between Independents and Presbyterians, by stating there was 'not a quarrel between Independencie and Presbyterie, which some would have'.[75] If this was not so, why was Cromwell a supporter of Self-Denial, for he would be removed from his own command by the strict observance of the word. By supporting, possibly even framing, the majority statement in the Ordinance, Cromwell had laid himself upon the altar of self sacrifice, it had been an example to all who were honest in their intent upon the public faith, or an act of sheer hypocrisy. The debate on Cromwell's integrity in Self-Denial has raged since his own time, yet in December 1644-January 1645 no one was in a position to know the future path the Ordinance would take, it is therefore dangerous to assess the situation with political hindsight.

It was against this backdrop of division that the New Model Army saw its birth. The Self Denying Ordinance had, by mid January, found increasing difficulties in the Lords, being rejected totally on 15 January 1645. At the same time that the ordinance (which would cleanse the army) was finding opposition in the Lords, the Commons pressed on with the formation of the army itself. The progress on the new army was due almost entirely to the vehemence of the Vane group, and must be weighed against the charges made against the Independents. That the Independent's would gain from a new army was not a forgone conclusion in January 1645, but it is certain from the sources available to conclude that the Vane, Cromwell, Hesilrige group, was more vehement in its pressure for change. The terms Presbyterian and Independent should not however be used to divide religious groups into pro and anti new army sub groups.[76]

On 21 January a major decision was passed when Sir Thomas Fairfax was named as General of the 'New Model'.[77] Fairfax was the son of Ferdinando Lord Fairfax, the commander of the Northern Association, who had fought an effective, if restricted war for Parliament in the north. Sir Thomas was a competent officer, of unquestionable loyalty, and he had not suffered from the jealousies within the other armies. Furthermore, Sir Thomas was a good choice in smoothing the nerves of the Lords, for he was not a member of the Commons which would have debarred him from service through the Self-Denial, but was close enough to the peerage via his father, to appear to be a continuation of Lordly privilege. It is interesting to note that Fairfax was not titled Lord General, for Essex was still the only commander to whom this title applied. The title Lord General is often wrongly given to Fairfax in 1645, it being in England used to denote the commander in chief of English Armies at a time when a member of the peerage alone was usually placed in such a position. Yet it is well to hypothesise upon the point, that Fairfax being not a Lord, could not be, and was not commissioned Lord General, but simply General.[78] Neither was Fairfax given military honour in the guise of public display, his appointment carrying no ceremonial act to set him apart. From the very beginning Fairfax was honoured for his deeds, treated with the most cordial of praise for his successes, and forever heralded the champion of Parliament for their salvation, yet at all times he took the part of soldier dutifully employed in the service of the nation, and not at any time a particular example of class power. This latter point set the new commander apart from Essex, who even in defeat had entered London in triumph bordering to almost Romanesque style.[79]

Writing of his employer, Joshua Sprigg gave a short history of Sir Thomas Fairfax's character and previous military experience:

> Sir THOMAS FAIRFAX, eldest Son of the Lord FAIRFAX, of *Denton* in the County of *York*: Martially disposed from his youth, Not finding action suitable to him in his own Country, (for through the great goodness and long suffering of God, *England* hath been a quiet habitation these 80 years) And there being imployment in *Holland*, he went over thither to enable himself in military experience: And upon his return into *England*, he marched into a most Noble and Martial family, taking to Wife one of the Daughters of that ever Renowned *General*, the Lord VERE. And thus the Reader may take notice, how not only his Extraction, Disposition, and Education bespake him for a *Souldier*, but his Contract also portended nothing less.[80]

Fairfax was therefore a fully trained soldier with an impeccable pedigree. That he served in a Dutch Army was a testimonial to the quality of his military education, but added to his ability on the field of war, must be added Fairfax's marriage into the Vere family. In classical terms, Sir Thomas had been blessed by

Mars the god of war, yet it was *Venus* who encompassed him into the heart of the military tradition in England. Perhaps in this respect it is worth consideration that by appointing Fairfax to command the new army, the Parliament had paid some small lip service to military tradition, through his own father and the point that Sir Thomas was the heir to the Fairfax peerage, this in addition to the military dowry accepted through his marriage. The Lords Vere had long been military champions and servants to the monarchy and peerage of England. Therefore, to some degree, Fairfax had the necessary breeding and connections to give the impression of the established class structure within military tradition. In direct comparison with the Earl of Essex, it is obviously difficult to draw social equilibrium with Sir Thomas Fairfax, but of those members of the gentry eligible for command after Self Denial, it is equally difficult to find anyone with similar qualification.

Continuing his *curriculum vitae* of Fairfax's personal history, Joshua Sprigg writes:

> Albeit, so far was he from congratulating such a condition of his Country, when he saw it like to need the exercise of his faculty, that he most sincerely offered the first attempts of his resolved Minde at the Altar of Peace. When at the King's first endeavours to raise a Guard for his own person, at York; apprehended then by those parts, and found quickly after to be the beginning of an Army: He was entrusted by his Country to present a Petition to His Majesty; the scope whereof was to beseech him to hearken to his Parliament, and not to take that course of raising Forces, he was then engaging in: which Petition the King refusing, he prest with that instance and intention, following the King so close therewith in the field, called Heyworth Moor, in the presence of 80, if not 100,000, people of the County (the like appearance was hardly ever seen in Yorkshire) so close, I say, til at last he tendred the same upon the pomel of his saddle. But finding no Propitiatory here, and seeing a War could not be avoided, he early paid the vowes of his Martial dedication. And so soon as these unhappy troubles brake forth, took a Commission under his Father, Ferdinando L.Fairfax, (whose timely appearance and gallant performances for his country in the North, deserves a story by it self;) And served the Parliament in lower Commands, then what Providence since hath adjudged his capacity and merit unto, making him now General of the Parliaments forces; to which trust and honour he was preferred upon no other grounds than the observation of his Valour, and all answerable abilities for the same, testified in many notable services done by him in the North, whilst he was yet in a lower sphere. And now how delightfully remarkable is it (as a most apt cadency of providence) if God shall make him, who was by the King rejected in his wild endeavours to prevent the troubles of the Land by a Petition, (then which he sought

nothing more) a most powerfull Instrument of restoring Peace thereunto by the sword.[81]

This picture of Fairfax as an instrument of god's judgement is both convenient and symbolic of Parliament's thinking. He is shown as first the peacemaker, yet upon his rejection, he is transformed into the King's bane. Sprigg does mention that his employer, Fairfax, was chosen 'upon no other grounds' than his valour and ability, suggesting that pedigree and patronage had nothing to do with his appointment. However, in a country where birth had held precedence over ability in every sphere of life, it must be said that England was most fortunate in having such a balance of lineage and ability in its new General.

Fairfax was to have under him a very considerable field army, in theory not weakened by garrison duty. For the first time since before Edgehill, the principle of the new army was to defeat the Royalists in the field. The actual regimental establishment strength of the New Model was to consist of the following:

Horse................6,000 men
Foot.................14,000 men
Dragoons.............1,000 men
TOTAL...............21,000 men

The Horse were to be in ten regiments of 600 men each, and sub divided in six Troops of one hundred. Twelve Regiments of Foot were to be formed with 'at least' 1,200 men to a regiment. Each regiment was to be sub-divided into ten Companies. The Dragoons were to form a single regiment in ten Companies.[82] If twenty-one thousand men could be put into the field as a single army, then Parliament could present against the King a force in line with the force at Marston Moor, and superior to that so dismally inadequate at Donnington.

At the same time, estimates of the cost of the new army were made and submitted to the Committee of Both Kingdoms:

The monthly charge of 14,000 foot in 12 regiments, and 6,000 horse in 10 regiments, with 1,000 dragoons in single companies, according to the last establishment.

The troopers at 2s. a day:

	£	s.	d.
The officers of a regiment of foot, £73 10s. 0d. For 12 such regiments............................	882	10	0

The officers of the colonel's company, £38 5s. 4d For 12 such companies……………………………..	459	4	0
The officers of a private company £31 19s. 4d. For 108 such companies……………………………… For 14,000 foot, at 8d. a day……………………..	3, 452 13, 066	8 13	0 4
TOTAL……………………………………………………..	17, 860	5	4
The officers of a regiment of foot, £61 2s. 8d. For 10 such regiments………………………....	611	6	8
The officers of the colonel's troop, £120 3s. 6d. For 10 such companies…………………………….	1, 201	13	4
The officers of a troop of carabineers £115 19s. 4d. For 50 such troops………………………………… For 6,000 troopers at 2s. a day………………….	5, 798 16, 800	6 0	8 0
TOTAL……………………………………………………..	24, 417	5	4

The officers of a company of dragoons, £58 2s. 0d. For 10 such companies………………………………… For 1, 000 dragoons at 18d. a day	581 2, 100	0 0	0 0
TOTAL……………………………………………………..	2, 681	0	0
GRAND TOTAL……………………………..…………..	44, 952	12	0

These estimates not only give a rough manpower cost, but also relate that the troopers in the new army were to be carabineer - armed with sword, pistol and carbine. This reduced the need for more than a single regiment of dragoons, because being armed with the heavier shot from the carbine, the horse could more easily confront the enemy foot, or contribute to sieges.[83] It is interesting that the term harquebusier was superposed by carabineer by the committee and indeed the Lords, leading to the false conclusion that troopers armed with carbines were unfashionable in England.[84] Many troopers being of a better class, already possessed their own sword, pistols and carbine from their Trained Band service, therefore there is no reason why for example Cromwell's troopers, in the main, needed to be issued with the latter.[85]

On 21 January, the House of Commons passed the order to establish the pay structure, and also desired that the names of the Lieutenant Colonels of Foot, and Majors of Horse were to be named. These officers were, in the seventeenth century, the recognised second-in-commands of their respective arms, and often in actual regimental command while their Colonels' were at headquarters or other administrative duties. Sir William Balfour also received the thanks of the House for his faithful service, and order was given to consider how officers laid aside by the reductions into the new army were to be best used in 'the west or elsewhere'.[86]

Such a force had not been seen in England before, as the army of 1642 had still been founded in privilege. The New Model Army on the other hand was a national standing army, raised and wholly maintained by state taxation, this provision in itself being revolutionary insomuch that loyalty of this army was solely in the nation and not in any individual liege or lord.[87]

The army structure had now established, and the General had been named. The question of ethical and religious standing was now to become the burning issue.

NOTES

[67] Richard Baxter, *Reliquiae Baxterianae, or Mr. Richard Baxter's narrative of the most memorable passages of his life and times* (1696) part I, p.47.

[68] For a wider debate see J. S. A. Adamson, 'The Baronial context of the English Civil War', *Transactions of the Royal Historical Society*, 5th series, 40 (1990) pp. 93-120.

[69] John Rushworth, *Historical collections of private passages of state* (7 vols., 1659-1701) part IV, vol. 1, p.3.

[70] Baxter, *Reliquiae Baxterianae*, part I, p.47.

[71] Ibid. p.98.

[72] Bulstrode Whitelocke, *Memorials of the English Affairs* (4 vols., Oxford, 1853) vol. 1, p.209.

[73] BL Thomason Tracts E.256[42], *Perfect Occurrences*, No.16 (22-29 November 1644) pp.2-3.

[74] A. N. B. Cotton, 'Cromwell and the Self Denying Ordinance', *History* 62 (1977) pp. 211-31, esp. p.221. Cotton rightly questioned Henry Walker as a reliable reporter, for he was known for his dubious and pointed embellishment of facts. Yet the story was never in a position to be denied or retracted, adding to its weight and acceptability at the time.

[75] BL Thomason Tracts E.21[3], *The Parliament Scout*, No.76 (28 November - 5 December 1644, pp.606-607.

[76] Sir William Waller was religiously a Presbyterian and later supported the Holles/Stapleton faction in Parliament. Yet in the matter of the new army he rigidly voted with the Independents.

[77] Rushworth, *Historical collections*, part IV, vol. 1, p.7.

[78] Ibid.

[79] After the defeat of Essex around Lostwithiel, he had made a formal entry into London with two regiments as guard d'honneur where he was greeted by the sheriffs of the City and the principal citizens. See L. C. Nagel, 'The militia of London, 1641-49' (Ph.D. thesis, University of London 1982) p.206.

[80] Joshua Sprigg, *Anglia rediviva; or England's recovery* (1647) pp.7-8.

[81] Ibid. pp.8-9.

[82] Dragoons were, at this time, considered to be mounted Foot and therefore in Companies not Troops.

[83] *C. S. P. Dom. 1644-45*, p.232.

[84] C. H. Firth and G. Davies, *The regimental history of Cromwell's army* (2 vols., Oxford, 1940) vol. 1, pp.2, 17, 19, totally reject the term harquebusier in the context of the English horse, but fail to accept the Anglicism into the term carabineer which survives to today.

[85] Hesilrige's troopers in Waller's army carried carbines from 1643 onwards.

[86] *C. S. P. Dom. 1644-45*, p. 265.

[87] Although the constitution of the nation would change with the Commonwealth, Protectorate and Restoration, at no time did the English Army pass back into single hands without the consent of a form of Parliament. Even during the rule of the Major-Generals, the Protector, could not raise men for his own benefit.

CHAPTER IV

THE TAKING OF THE COVENANT

Simultaneously with these debates on the army, the Houses had been sounding out the possibilities of peace on their terms and had despatched a team of negotiators to meet royal commissioners at Uxbridge. By February 1645 the Uxbridge proposals had reached a point where, in reality, the Commissioners for both sides knew that a peaceful settlement to the nations misery was even more improbable than they had expected. It was by now imperative that Parliament ended its futile inner bickering, and seriously prepared its new army for the field.

There is at least some evidence that in the realization by the City that Uxbridge was a sham, the loyal merchants of London fearing a sudden attack by Prince Rupert, or even the King himself, finally lost both their nerve and patience with their political allies, and turned the financial screw sufficiently tight to demand protective action.[88] The Scottish commissioner Loudoun most eloquently summed up this feeling of impending failure at Uxbridge, by reporting that, 'it was then high time for the watchmen to give the alarm to the people, and warn them of their danger, lest they should be deceived with the vain hope of peace'.[89]

On 11 February the report of the joint Conference for the Ordinance for remodelling the army, and to place it under the command of Sir Thomas Fairfax, reached the House of Lords. This date can and should be seen as a watershed in the progress of the reorganisation of the army. From this report stage, a new, more determined approach can be seen to come from the more hawkish members of the Commons, who began to press for substantial changes in the Ordinance conditions governing the administration of the forces. At the same time, the conditions were delivered by John Lisle, a member of the Commons House, to a reconvened Conference meeting in the Painted Chamber. The substance of the alterations to the Lords original proposals, were recorded at some length, and are worth greater examination:

> My Lords,
> The First Alteration is in your Lordships First Proviso, which I shall First read: [Provided also, That all Colonels, Lieutenant Colonels, Serjeant Majors and Captains, that shall be employed in this Army, may be such as shall be nominated, appointed, and approved of by both Houses of Parliament.]
> The House of Commons desire it may be thus amended: [Provided also, That the Commander in Chief, nominated in this Ordinance, shall have Power to nominate all Colonels, Lieutenant Colonels, Serjeant

Majors, and Captains, out of the Forces under the Command of the Parliament.][90]

The proposal that Fairfax, and not Parliament itself, should nominate his own officers had a twofold effect. It would simplify the reduction of existing regiments by enabling Fairfax to take officers and men en bloc and at his own discretion, but also attempted to disseminate the power of the Lords, and overcome the delaying tactics of Essex's supporters in the upper chamber. Lisle explained the need to grant Fairfax this power thus:

> They [the Commons] do conceive, that, by giving of this Power to the Commander in Chief to nominate their Officers, will more oblige his Officers, and better enable him to carry on the Work. They do conceive, that there can be no Inconvenience in giving him this Power of nominating, in regard both Houses of Parliament are to approve of the Officers nominated by the Commander in Chief.
> And, they do observe, that this Power of nominating Officers is no unusual or extraordinary Power; for this Power of nominating is granted constantly, and usually to every Commander in Chief.[91]

True, when Essex had received his commission in 1642 he had received with it the power to nominate his own officers, and had passed down the power to each Colonel, almost without exception, who would nominate officers to serve under him. This practice had been modified in the restructuring of Essex's army in 1643 to give Parliament a greater control over its officer corps, but not until the first Committee of Both Kingdoms did Essex find any significant changes to the autonomous nature of his commission. Yet if Essex had been his own master in matters military in the first year of the war, this was also before the taking of the Covenant, and the question of the limitation set by this oath was still to come in the new army.

In the Painted Chamber the Conference members continued to hear the arguments to the Lords proposals, related in ever increasing depth by John Lisle:

> My Lords,
> They do intend principally, that these Officers shall be nominated by the Commander in Chief, out of any the Forces under Command of Parliament, whether out of the Lord General's Army, or that under Command of the Earl of Manchester, or Sir William Waller; and if this Clause had not been inserted, we could not have nominated any of those Officers.
> And the Clause is more general, because the Commander in Chief will have the greater Lattitude, if he desire it, to nominate such Persons for Officers as he shall think faithful, and fittest for Service.[92]

Even at this late date, Essex was diplomatically referred to as the Lord General with his own army still to command. The Commons were theoretically correct in ascertaining that neither House could nominate officers from the three armies designated for reduction into the new force, whilst Essex held the commission over the militia. But this legal definition of Essex's power, could, and would be, sidestepped with self induced amnesia by the Commons, and Fairfax's ultimate willingness to accept command of this spectral force of a would-be army. The actual willingness of Fairfax to serve might not in fact have been so cut and dried, for he later recalled, 'Had I not been urged by persuasion of my nearest friends, I should have refused so great a change.'[93] The taking of the Covenant held no fears for Sir Thomas, for in religion he was both moderate and conservative in the puritan way, yet it is obvious that he could see the fundamental changes incorporated into the concept of the new army.

The question of the religious Covenant could not however be so conveniently overlooked, and Mr. Lisle reintroduced it into the debate:

> Your Lordships Second Proviso is thus.
> [Provided further, That all Commanders, Officers, and Common Soldiers, that shall be employed in this Army, shall take the Solemn League and Covenant of both Kingdoms, within Twenty Days next after Publication thereof; and shall submit to the Form of Church Government that is already voted by both Houses of Parliament.]
> My Lords, the House of Commons desire it may be thus ammended.
> [Provided further, That all Commanders and Officers, that shall be employed in this Army, and to be approved of by both Houses of Parliament, as aforesaid, shall take the National League and Covenant of both Kingdoms, within Twenty Days next after Approbation; and all other Officers, to be employed as aforesaid, shall take the said Solemn League and Covenant within Twenty Days after they shall be listed in the said Army; And be it Ordained. That all the Common Soldiers of this Army shall likewise take the same, at such Time, and in such Manner, as shall be in that Behalf directed by both Houses of Parliament.][94]

To understand fully the reasons for such subtle alterations, a closer examination of the Covenant is needed. When on 25 September 1643 the House of Commons took the oath of the Solemn League and Covenant, it laid the foundation not only for the accommodation of Scottish military assistance in England, but accepted Presbyterianism as the one religion of church and state. In so doing, it introduced into the framework of the military hierarchy, a political schism which far outweighed any religious implication and confirmed the differences between the religious Presbyterians and the sectarian Independents in both Parliament and army.

The Covenant placed the King, in his absence, at the head of the joint Anglo-Scottish church, at the command of God and to further the cause of 'our lord Jesus Christ', but also within a constitutional framework of King, Church and State. The sections of the oath are worthy of greater investigation, for they relay within their eloquence, a strict understanding of the new Anglo-Scottish concept of the states' role in the reformed and layered society, and the difficulties this form of worship would bring in moulding the army of 1645. The Covenant opens:

> We noblemen, barons, knights, gentlemen, citizens, burgesses, ministers of the Gospel, and commons of all sorts in the kingdoms of England, Scotland and Ireland, by the providence of God living under one King, and being of one reformed religion; having before our eyes the glory of God, and the advancement of the kingdom of our Lord and Saviour Jesus Christ, the honour and happiness of the King's Majesty and his posterity, and the true public liberty, safety and peace of the kingdoms, wherein every one's private condition is included; and calling to mind the treacherous and bloody plots, conspiracies, attempts and practices of the enemies of God against the true religion and professors thereof in all places, especially in these three kingdoms, ever since the reformation of religion; and how much their rage, power and presumption are of late, and at this time increased and exercised, whereof the deplorable estate of the Church and kingdom of Ireland, the distressed estate of the Church and kingdom of England, and the dangerous estate of the Church and kingdom of Scotland, are present and public testimonies. We have [now at last] after other means of supplication, remonstrance, protestations and sufferings, for the preservation of ourselves and our religion from utter ruin and destruction, according to the commendable practice of these kingdoms in former times, and the example of God's people in other nations, after mature deliberation, resolved and determined to enter into a mutual and solemn league and covenant, wherein we all subscribe, and each one of us for himself, with our hands lifted up to the most high God, do swear.[95]

The essence of this opening reaffirmation of their religious cause, also committed their loyalty to the King; or at least a King who would head and defend the post reformation church in its Presbyterian form, and by so doing continued the myth of a King kept from his true chosen path by evil counsel. The oath, although not a formal constitution, continued to claim Scotland to be a kingdom within its own government, so was not a true act of national Union except in the unification of the soul. The oath in this form subcutaneously laid claim to Ireland through the triple crown of the realm. Those accepting the oath were engaged to defend the religious and state doctrine lain down by the Covenant, and the progressive steps of the form it took, were retrospectively, a bitter pill for many English soldiers to take.

The first part of the oath repeats the appeal to church reform and unification to the Scottish system, and continues in section two by removing church government by papal means, 'that is...by Archbishops, Bishops, their Chancellors and Commissaries, Deans, Deans and Chapters, Archdeacons, and all other ecclesiastical officers depending on that hierarchy'.[96]

By parts three and four however, the true purpose of this Covenant, and the reason for the English Parliament's interest in the Anglo-Scottish alliance, reveals itself:

> We shall with the same sincerity, reality and constancy, in our several vocations, endeavour with our estates and lives mutually to preserve the rights and privileges of the Parliaments, and the liberties of the kingdoms, and to preserve and defend the King's Majesty's person and authority, in the preservation and defence of the true religion and liberties of the kingdoms, that the world may bear witness with our consciences of our loyalty, and that we have no thoughts or intentions to diminish His Majesty's just power and greatness.

The last sub-section of this third part, repeating an oath of loyalty to the King, did not alter the fact that to replace the Anglican religion in the form of Episcopacy, with Anglo-Scottish presbyterianism, would in itself reduce his power. But it was the opening lines with the pledge to 'endeavour with our estates and lives' which committed Scottish manpower to preserve the Covenant in England, and the King's person wherever he might be.

In part four the oath moved on to alienate the King politically from those who supported him and configured his cause in terms of military power or political advocacy:

> We shall also with all faithfulness endeavour the discovery of all such as have been or shall be incendiaries, malignants or evil instruments, by hindering the reformation of religion, dividing the King from his people, or one of the kingdoms from another, or making any faction or parties amongst the people, contrary to the league and covenant, that they may be brought to public trial and receive condign punishment, as the degree of their offences shall require or deserve, or the supreme judicatories of both kingdoms respectively, or others having power from them for that effect, shall judge convenient.[97]

The above needs little explanation, it condemned such men as Ralph Hopton as war criminals, or 'malignants and evil instruments', but it also allowed for the service of Scottish troops in England, English troops in Scotland if conditions demanded peace keeping in defence of kingdom and religion, and moreover, committed both kingdoms to preserve the reformation and kingdom by force of arms in Ireland. A full understanding of the wording of parts three and four of the Covenant is vital if an understanding of the complexities of the moulding of the New Model Army is to be gained. They constitute the commission which governed all soldiers serving in England from the time of its inauguration, and opens the door to many absorbing questions in forming the consensus of the new army.

It is necessary to understand when dealing with the debates between both tiers of the English Parliament and the Scots, that the Covenant, by its binding of the national League against the enemies of the presbyterians, had transformed the military machine into a multilateral force from the unilateral force of 1642. The new army of 1645, at least in theory, was by the Covenant forced to adopt the oath if it was not to break the alliance. It would, if to conform to the League's criteria, be in effect pre-moulded in a form of the Scottish presbyter's choice, because it could not, without breaking free of the oath, form an army to act against the King.

Part five of the detailed oath, continued the propaganda war by strengthening the presbyterian right to fashion the church state in a self likeness, but the final section, part six, returned to a more politically motivated theme, saying,

> We shall also, according to our places and callings, in this common cause of religion, liberty and peace of the Kingdom, assist and defend all those that enter into this league and covenant, in the maintaining and pursuing thereof; and shall not suffer ourselves, directly or indirectly, by whatever combination, persuasion or terror, to be divided and withdrawn from this blessed union and conjunction, whether to make defection to the contrary part, or give ourselves to a detestable indifferency or neutrality in this cause, which so much concerneth the glory of God, the good of the kingdoms, and the honour of the King; but shall all the days of our lives zealously and constantly continue therein, against all opposition, and promote the same according to our power, against all lets and impediments whatsoever; and what we are not able ourselves to suppress or overcome we shall reveal and make known, that it may be timely prevented or removed; all which we shall do as in the sight of God.[98]

This was the inexorable part which bound the people to the Covenant, and the Covenant itself together. At the same time, it bound the individual thus commissioned through God and state, to abide by the stringent presbyterian doctrine, and the state's authority contained within its words. Although claiming to set at liberty the individual soul, the Covenant set restrictions to the physical body in its freedom of choice, and conscience of the mind. Many supporters of liberty and freedom of religion and political conscience could not, for this reason alone, take the Covenant oath and serve in the army.

It would seem that the requirement to swear the oath was not universally strong within the Parliamentary armies, the Associations being less insistent upon it at least in its earliest days. The Lord General Essex on the other hand, being the direct authority of Parliament's military power, could not ignore the Covenant, or as some maintained, it gave a salve to the inextricable stance of fighting for 'King and Parliament', by replacing these false words with those 'For King, Religion and Covenant.'[99]

One such officer to fall victim of Essex's rigid adherence to the Covenant was John Lilburne, who wrote:

> ...at my coming home, some of them that were no mean ones, proffered my wife a place of honour and profit for me, then reputed worth about 1000l.per annum; which I conscientiously scorned and slighted, professing unto my wife, to her extraordinary grief, that I must rather fight for 8 pence a day, till I see the liberties and peace of England setled, then set me down in a rich place for mine own advantage, in the midst of so many grand distractions of my native Country as then possessed it; and so I left old Essex, that had been so generous unto me in giving me almost 300l. ready money at my deliverance, as Colonel Fleetwood and Colonel Harrison very well know. But him for all that [I say] I left, for his persecuting for non-taking the Covenant, and down to Lincolnshire I posted, to my then two Darlings and familiar Friends, Manchester and Cromwel, where I engaged heartily'.[100]

Poor Mrs. Lilburne, she never once benefited from her radical husband's populist stances, but perhaps John Lilburne was more married to his natural destiny than to his wife. From Lilburne's statement, it is evident that he and others had sought freedom from the restrictions of the Covenant, and this freedom was rediscovered in the Eastern Association. If one therefore considers that the Associations were to provide large numbers of trained soldiers; trained one must add in religious sectarianism as well as the art military, then it is not difficult to accept that the new army stood or fell by the Covenant.

Returning to the details given by John Lisle on 11 February, he then began to explain the reasons for the Commons amendments:

> My Lords, for the First Amendment, they present these Reasons.
> They do observe, that in your Lordships Proviso the Time is not certainly expressed; for your Lordships Expression is ['they shall take the Covenant within Twenty Days after Publication thereof']; and therefore, to make it more certain, they have added this Clause, ['that they shall take the Covenant within Twenty Days next after the Approbation of

both Houses.'] And not only the Commanders and Officers approved by your Lordships, but all others, shall take the Covenant, and therefore it is added, ['that all Officers whatsoever are to take the Covenant within Twenty Days after they shall be listed in the said Army.']

Not to have added these words could, in the fullness of time and with the usual manpower wastage on campaign, have lead to an army officered by Covenant and non-Covenant commanders. Turning to the Common soldiers however, the amendments touch upon a sore point:

> My Lords, concerning the Amendment for the Common Soldiers in Point of Time, they have made it on this Ground. They do conceive that you may be inforced to press some Soldiers, to serve you in this new Model; and if you should be inforced thereunto, these Soldiers may make their Excuse of not serving the Parliament, on Pretence they cannot take the Covenant.
> They do observe, besides, that the recruiting of the Army is uncertain; they do not know the certain Time, and think it not convenient that, before the Army be recruited, they should set down any certain Time for the Common Soldiers; but that it be referred to the Wisdom of both Houses.[101]

The members of Commons were correct in their prognostic interpretation of the difficulty in binding the rank and file to the Covenant. If the army was to have liberty of conscience, and yet not liberty of body from the press gang, no soldier could be forced to serve against his conscience though his body was pressed to do so. The pressing of soldiers had been used in the opening years of the war to principally augment depleted regiments upon the march, but manpower to the armies had been dramatically drained by battle, and moreover, disease, and by 1645 the pressing of unwilling recruits was becoming increasingly necessary.[102] The realization on 11 February, that the three armies of Essex, Manchester and Waller could not alone supply the whole infantry requirement for the new army, also highlighted the structural difficulty in maintaining the adherence to the Covenant in a rigid form. The situation was also enhanced by exemptions to being pressed for members of Trained Bands and the County Militias of Horse, which had existed since Essex's original commission in 1642.[103]

The Impressment Act of 13 February 1642 had been passed with some urgency to raise an army supposedly for service in Ireland, although all knew its real purpose. In its order for the Array of 'soldiers, gunners and chirugeons', exceptions were made as follows:

> Provided always that this Act shall not extend to the pressing of any clergyman, or any scholars or students or privileged persons of either of the Universities, Inns of Court or Chancery, or any of the trained bands of this realm, or to the pressing of any other person who was rated towards the payment of the last subsidies, or that should be rated or taxed towards the payment of any subsidies hereafter to be granted before the time of such impressing, or to the eldest son of any person who is or shall be before the time of such impressing rated in the subsidy book at £3 lands or £5 goods, or to any person of the rank or degree of an esquire or upwards, or to the son of any such person of the said rank or degree, or of the widow of any such person, or to any person under the age of eighteen or above the age of threescore years, or to any mariners, seamen or fishermen.[104]

It is evident from this section of the Act that it was the 'poorer sorts' of people who were pressed for military service during 1642-45. Those with an 'interest in this nation' to use Henry Ireton's term from 1647 could not be pressed into service.[105] This did not prevent the members of the Universities or Inns of Court from volunteering for military service, many like Edmund Ludlow or Charles Fleetwood had joined the Life-guard from the Inns of Court, and Charles D'Oyley had been a student in Oxford.[106] Henry Ireton was exempt from impressment by the social exemption through his widowed mother, but still served from the outset, whereas his brothers chose to use the clause to farm the family's Nottinghamshire estate. However, with the exception of servants to members of the Lords and Commons, and 'the inhabitants of the Isle of Wight, or of the Isle of Anglesey, or of any of the Cinque Ports, or members thereof', other menial servants were subject to impressment.[107] It is likely that Nehemiah Wharton whose graphic account of the 1642 campaign relays so much of the common soldier's life, was so taken for service from his London master, to serve in Denzil Holles's regiment.[108] Although the Act was for but a one year period, it was renewed by successive sittings of both Houses, and was in force in a relatively unchanged form for the new army.

Obviously few servants, and possibly even fewer servants with their masters blessing, would take the Covenant if taken for impressment. The fact that it was the lower classes who were pressed, also questioned their ability to be held to an oath, because only those who understood to what they swore, could be held responsible to their word, which accounted for the numbers of impressed men who deserted.[109] It is estimated that more than half of those pressed to supplement the garrison at Newport Pagnell during 1645 deserted.[110] Despite the passing of a renewed act for martial law in various garrison towns, including Aylesbury and Newport Pagnell, the rate of desertions was never brought under control in the non-remodelled regiments, and the need to regulate for a standing army can be emphasised by this fact. The general injustice of this early form of conscription was nonetheless, dealt with by William Walwyn in the Remonstrance of 1646 when he writes:

> We entreat you to consider what difference there is between binding a man to an oar, as a galley-slave in Turkey or Algiers, and pressing men to serve in your war. To surprise a man on the sudden, force him from trade, from his dear parents, wife or children, against inclination...to fight for a cause he understands not, and in company of such as he hath no comfort to be withal, for pay that will scarce give him sustenance, and if he live, to return to a lost trade or beggary, or not much better. If any tyranny or cruelty exceed this, it must be worse than that of a Turkish galley-slave...The Hollanders, our provident neighbours, have no such cruelties...yet they want no men.[111]

This doctrine is not pacifist in its concept, for many of Walwyn's persuasion would fight upon their own conscious terms, but it does question the whole conceptual essence of forcing the nation's will upon an unwilling participant. Had the Covenant been forced upon the common soldiers therefore, it would have provided an open and shut clause for conscientious objection, and would have led unquestionably to a weakened recruitment availability.

By the close of Mr. Lisle's speech to the assembled conference, there can have been few, if any, Peers who did not realise that the Commons were preparing to press not only soldiers, but their own will upon the upper House, to shape this army in a form with less lordly grace yet more practical purpose. The 11 February had seen rapid progress in the moulding of the army, for by proposing that the officers should be chosen by Fairfax; the Lords and Commons retaining a veto upon the lists, the physical work of reducing old armies to assemble new could be underway within weeks, and the new army be ready before the Spring offensive was under way.

The Commons had introduced problems into the question of the Covenant, and it was widely known that Cromwell and the younger Henry Vane were hostile to the taking of the Covenant by this army at all, but if it had to preserve the alliance with Scotland and if the Solemn League was the agreement by which the treaty stood, then it was also bound by expedience towards a military reorganisation and a desire from more sectarian quarters not to become totally subservient to the Scottish Kirk.

The occasionally outspoken newspaper, *The Scottish Dove*, recognised the problems of sectarian division within the Commons, and was at this time highly sceptical of the Parliaments' true faith in the Covenant, and not without good cause. The problems recently seen in the Eastern Association between Cromwell and Crawford had been to no small degree, due to sectarianism within the ranks of Cromwell's supporters. Cromwell's newly strengthened working relationship with the younger Vane, and the deterioration between the moderates and war party since the death of John Pym, had done nothing to convince the Scots of the sincerity of the English adherence to the Scottish religious and political culture represented by the Covenant. On the other hand, a month later, the Royalist newspaper *Mercurius Aulicus* was more critical of the Commons' plans for the new army, informing its readership that,

> their Solemn League and Covenant is now only tendered to good consciences whom they hope will refuse it, and waived as often as any sort of Rebels pretend to stumble at it...they have now granted that their new General, Sir Thomas Fairfax, and all the officers of their intended Army, shall not have the Covenant prest upon them, and yet their commissioners at Uxbridge will have no peace unless His Sacred Majesty sweare to this covenant and injoyne it to all his subjects.[112]

With this stance by *Mercurius Aulicus* in mind, it is not surprising that the restrictions of the Covenant would quickly fall victim to political expedience. The oath which bound the Alliance together would, if its critics were correct, falter, and the new army would break its chains of conscience.

NOTES

[88] Lawrence Kaplan, *Politics and Religion during the English Revolution: the Scots and the Long Parliament 1643-1645* (New York, 1976) p.110. Kaplan maintains that Sir Henry Vane Junior had strong links with the economic base of the city, a point proven by Violet Rowe in her excellent 'administrative biography' of the younger Vane.

[89] BL Thomason Tracts, E.273[3] *The London Post*, No. 27 (11 March 1645).

[90] *L.J.* vii, p.189.

[91] Ibid. p.190.

[92] Ibid.

[93] Mark Kishlansky, *The rise of the New Model Army* (Cambridge, 1979) pp.40-41.

[94] *L.J.* vii pp.189-190.

[95] John Rushworth, *Historical collections of private passages of state* (7 vols., 1659-1701) part III, vol. 2, p.478; 'The Solemn League and Covenant', 25 September 1643, printed in *The constitutional documents of the Puritan Revolution 1625-1660*, ed. S. R. Gardiner (Revised edition, Oxford, 1906) pp.267-8.

[96] 'The Solemn League and Covenant' in Gardiner (ed.), *Constitutional documents*, p.269.

[97] Ibid. pp.270-1.

[98] Ibid.

[99] All banners carried by Scottish Covenant infantry regiments from 1643-51 invariably carried as a motto the commitment, 'For God, Religion, King and Covenant', thereby committing their forces to preserve by military means, up to and including those of David

Leslie in 1651, any King of the duel Anglo-Scottish throne, who would adhere to the national League and Covenant.

[100] John Lilburne, *The legall fundamentall liberties of the people of England* (1649), extracts printed in *The Leveller Tracts, 1647-1653*, eds. William Haller and Geoffrey Davies (second edition, Gloucester, Mass., 1964) pp. 399-449, quote on p.407. Lilburne had been taken prisoner fighting for the Parliamentary cause at Brentford on 12 November 1642, and became a popular, almost folk hero in London for his defiance of the King's authority over him, *Dictionary of National Biography*, q.v. Lilburne, John.

[101] *L.J.* vii p.190.

[102] Kishlansky maintains, quite correctly, that the Home Counties around London had primarily provided the manpower for the Earl of Essex's 1643-45 army, *The rise of the New Model Army*, p.22. However, he possibly fails to fully acknowledge the value of the London Trained Bands in supplementing the infantry requirement.

[103] 'The Impressment Act', 13 February 1642, printed in Gardiner (ed.), *Constitutional documents*, pp.243-5.

[104] Ibid. p. 244.

[105] The Putney Debates, in *The Clarke Papers*, ed. C. H. Firth (Camden Society publications, 4 vols., 1891-1901) vol. 1 (1891) pp.226-440. During the debates, Ireton laid great emphasis upon those with an 'interest' in the nation. It is clear that only those with land or taxable income should, in Ireton's view, receive rights of the political franchise. This view is then in complete accordance with the Impressment Act of February 1642.

[106] C. H. Firth and G. Davies, *The regimental history of Cromwell's army* (2 vols., Oxford, 1940) vol. 1, p.42.

[107] 'The Impressment Act', Gardiner (ed.), *Constitutional documents*, p.245.

[108] See *The Edgehill campaign and the letters of Nehemiah Wharton*, ed. S. Peachey (Leigh-on-Sea 1989).

[109] The diary and payroll records of Captain Samuel Birch's company in 1648 show large numbers of desertions during the active periods in the field, but comparatively stable in number during garrison duty.

[110] *The letter books, 1644-45, of Sir Samuel Luke, Parliamentary General of Newport Pagnell*, ed. H. G. Tibbut (Publications of the Bedfordshire Historical Record Society 42; HMC joint publications, 4, 1963) pp.258, 413, 422, 548.

[111] Quoted at length in H. N. Brailsford, *The Levellers and the English Revolution*, ed. Christopher Hill, (Spokesman University Paperbacks, no. 14, Nottingham, 1976) p.101. Brailsford is particularly strong in dealing with such problems of the poorer classes, perhaps identifying to no small extent between the pressed soldier of 1642-46 and the conscripted Tommy of 1914-18. The Leninist message to the Eastern Front conscripts in pre-revolutionary Russia, had differed little to that relayed by Walwyn two hundred and fifty years earlier.

[112] BL Thomason Tracts E.271 [4] *Mercurius Aulicus*, (2-9 February 1644). *The Scottish Dove*, a highly outspoken news sheet supporting Parliament's cause, wrote similarly in its sixty-eight issue (31 Jan-7 February issue, BL Thomason Tracts E.269[3]), and relates the presbyterian feeling before the 11 February amendments.

Cromwell's Soldiers

The Taking Of The Covenant

CHAPTER V
GENERAL FAIRFAX

The debate on the Covenant continued on its way, but this single problem did not prevent the structure of the army from continuing, simply the method upon which the ideological basis of that army would be framed. When the final analysis is made it would be the ideology behind the new army which would govern the way it progressed.

In the third week of February, Sir Thomas Fairfax arrived in London in response to an express command from Parliament. With him came a corps of loyal Colonels, all of whom had loyally served the Parliament, they included Sir William Constable, Colonel Rigby, Colonel Sandys and Colonel Alured, none of whom could expect senior commissions in this new army. The next day, being 19 February, four members of the Commons went to Fairfax, and accompanied him to the House, where Sir Arthur Hesilrige introduced him to the Speaker and the members there assembled. A chair had been placed for their new champion, but modestly Fairfax refused it 'and stood bare' while the Speaker commanded him thus:

> That that House out of great Experience and Confidence they had of his Valour, Conduct, and Fidelity, had thought fit to confer the great Trust of Commanding their Armies in Chief upon him: And giving him Thanks in the Name of the House for his many and great Services past, encouraged him to go on as he had begun, assuring him of the Care and Protection of the Parliament in the discharge of that weighty Trust which the Kingdom reposed in him.[113]

That Fairfax had humbled himself before the Commons by standing throughout the Speaker's address, bound him to them as sure as any verbal reply would have done. The Commons and Lords could, without doubting his integrity, agree that their choice of General was a sound one, yet at this stage it was not thought fit to allow their new champion to commission his own officers. Although Fairfax was desired to submit lists, the question of who was to receive employment in the new army was for the joint agreement of both houses, not for the military establishment to administer to its own shape. In this position can be identified the new power of the Commons in its freedom against the executive power of King and Peerage, therefore the moulding of the New Model Army was made in a perfectly new form, untainted by the court or the power of privilege. This was certainly the vision of many of the Independent faction, who had long put 'godly interest' before the privilege of birth.

The term 'godly' had fast become the acid test for the Independent group, it was associated with purity of religion, but more and more was coming to mean purity of conscience. It is no repetition to assimilate the word 'godly' with Cromwell's officer who knew what he loved, and loved what he knew, for to be Cromwellian was to hold a vision not only of puritan grace, but the sense of where the nation's destiny was to lead.

If Fairfax was honest and acceptable to all parties in Parliament, the opinion of Richard Baxter suggested another reason for his appointment:

> When this was done, the next Question was, Who should be Lord General, and what new Officers should be put in, or old ones continued? And here the Policy of Vane and Cromwell did its best: For General they chose Sir *Thomas Fairfax*, who had been in the Wars beyond Sea, and had fought valiantly in Yorkshire for the Parliament, though he was overpowered by the Earl of *Newcastle's* Numbers. This man was chosen because they supposed to find him a Man of no quickness of Parts, of no Elocution, of no suspicious plotting Wit, and therefore One that *Cromwell* could make use of at his pleasure. And he was acceptable to sober Men, because he was Religious, Faithful, Valiant, and of a grave, sober, resolved Disposition; very fit for Execution, and neither too Great nor too Cunning to be Commanded by the Parliament.[114]

Baxter's supposition that Vane and Cromwell chose the General in the guise of Fairfax was not too unreasonable, for he was all the things the former said about him. The strength of Sir Thomas Fairfax in years to come was his ability to tread a centre course, but whether this was through choice or a lack of willingness to commit himself to a pseudo-religious political stance is difficult to ascertain. Of course unlike Manchester and Crawford, Fairfax had distinguished himself at Marston Moor, and Cromwell knew him to be a good businesslike officer. Lack of fine speech is of no detraction for a soldier, and it must be said that little of this social grace is to be found in his letters. However, a quote attributed to the King in calling Fairfax "their new brutish general" is hardly worthy of consideration. Sir Thomas Fairfax appears to have been nothing more nor less than a rather typical Yorkshireman, never using more words than the subject required simply to exercise the tongue. Yet he was the sort of commander soldiers would follow, he would stand with his men through the hottest assault, encourage them in battle, and through his gallantry inspire the very best from his troops.

At the same time that Fairfax was appointed General of the new army, Philip Skippon was made Sergeant-Major-General to command the foot. This latter appointment confirmed that Skippon was supreme in both ability and affection of Parliament, for Crawford and Holborne had been disregarded in their choice. Skippon had been the darling of London since his show of loyalty to the 'five members' in 1642, and his command of the Trained Bands in the stand off at Turnham Green in November of that year. In this position he had served Parliament faithfully, later leading Essex's foot and dealing with considerable honour in the surrender at Lostwithiel. Skippon was the obvious choice for Sergeant-Major-General, politically being acceptable to Essex and the Independents alike, because if any officer serving in England could be considered apolitical, it was this reliable soldier.

In January 1645 an alarm to possible Royalist incursion into the south and a strengthening in the west, caused Parliament to order Sir William Waller as General, with Oliver Cromwell as his Lieutenant-General of Horse, to march southward as a defensive force. This story has little direct bearing upon the New Model, but indirectly highlights the problems which needed to be overcome before the new army could be moulded. Upon receiving their order to march, the horse commanded by Balfour in Essex's army, mutinied, and said they would not serve under Sir William Waller. The newspaper *Perfect Passages* relates the situation,

> they have received divers complaints against some of our Officers of Horse, wherein was reckoned up, such and such horse to plunder, and among the country people, others to mutinie at the motion of going with Sir William Waller, in his advance for the West, others complaining that they wanted saddles, and other things, as pistols and the like, and for Colonel Cromwell's souldiers, it was informed, that in what posture soever they were, that (were it at midnight) they were alwaies ready to obey any Ordinance of Parliament, and that there was none of them known to do the least wrong by plunder, or any abuse to any country people where they came, but were very orderly, and paid onely for what they had, and were ready to advance with Sir William Waller. Now let us consider, and lay these things to heart, for (as a person of note then said) these are the men (meaning the Independents, as we call them) that the other day many would have had to have been turned out of the Army, you see how they submit to authority, and to be obedient, and how orderly they are, and readie to do service for the publicke. And we had more need to fall downe upon our knees to pray them to stay in the Army, then to go about to put them out; and truly for my part, I never tooke them for friends that went about to make divisions between us and them in the Army, for however they differ in something in judgement from us, yet we cannot be ignorant that they have been as active for the Parliament both in purse, person and prayers, as any in the Kingdome.

Cromwell's troopers, therefore, were acknowledged to be highly disciplined and motivated by a personal understanding that maintained their sincerity to the cause common to all factions of the Parliamentary ethic. The defence of the Independent's by *Perfect Passages* is a direct reference to the argument over the

Covenant, but takes the stance that these soldiers were no less, and possibly more, faithful than the Presbyterians.[115]

The Eastern Association horse set out with Waller, but through being required by the Parliament and Army Committee, Cromwell did not go with them. By late February nevertheless, Cromwell's troopers were themselves in mutiny. Parliament had ordered them a fortnight's pay on 5 February to ease their marching, yet the paymasters in the Eastern Association had still not converted the order into hard cash on the 20th when the order had to be repeated.[116] On 22 February the Speaker received letters from Waller which,

> relating how the Earl of Manchester's forces did now refuse to go with Sir William Waller westward as well as the Lord General's, because they had hitherto received no money as was promised, and yet they were ordered to carry with them four days provision for horse and men, by reason the countries into which they were to go were so wasted that they were like a wilderness, and had no provision for them; besides all this they were to be recruited and furnished with pistols which they were in great want of.[117]

A second letter of the 22nd, also from Waller relates that '700 of Cromwells horse, do what their officers could, had left them that afternoon, and were returned with colours to former quarters, pretending they had received great injuries and affronts.'[118] D'Ewes, who records the report in the Commons, is not clear if Waller was referring to Cromwell's own regiment, or uses to term 'Cromwells Horse' to mean the troopers of the Eastern Association, rather than his own Western or Southern forces.

To quell any disobedience in the Eastern Association, the Commons released Cromwell from his Parliamentary duties. The order was made on 27 February, and £1,000 was voted as payment to both the Lieutenant-General and his own regiment, which further order to take Ireton's and two other troops of his double regiment, currently quartered at Henly, with him.[119]

In a direct relation to Cromwell's own speech when proposing Self Denial, the lack of care by Parliament, not the sincerity of his men, had led to these mutinies. Cromwell had said,

> I am not of the mind that the calling of the members to sit in Parliament will break, or scatter our armies, I can speak this for my own soldiers, that they look not upon me, but upon you, and for you they will fight, and live and die in your cause; and if others be of that mind that they are of, you need not fear them; they do not idolize me, but look upon the cause they fight for; you may lay upon them what commands you please, they will obey your commands in that cause they fight for.[120]

Yet through the inadequacies of pay and provision by Parliament, even Cromwell's troopers had mutinied. Gathering the three troops from Henley, Cromwell rode to join Waller who was already at Portsmouth. A report of 17-24 March in the *Perfect Diurnal* tells of Cromwell's action:

> A letter came this day to the House of Commons from Colonel Cromwell, a worthy and valiant member thereof, and one of the saviours (as God hath miraculously manifested him to be, of this Israel), informing the House, that since his coming to his regiment, the carriage of it hath been very obedient and respectful to him, and valiant, a good testimony whereof they gave upon the late service against Long's regiment [12 March]; and for their late mutinous carriage to the Parliament, they had expressed their hearty sorrow for the same, and had desired him to send their most humble petition to the House, that they might be again received into their former favour, and have their pardon fully for their late offence; and for the quite removing the cloud of jealousy over them, they doubted not to clear it by valiant testimony, as occasion should require. The House upon reading the same, received much satisfaction, and accepted it affectionately from them.[121]

The statement belied Cromwell's earlier statement on Self Denial, for this clearly exemplifies that although almost blasphemous in its concept, his soldiers were as much disciples to the man as subservient to the cause. To muster Cromwell's regiment an advertisement appeared in *The Moderate Intelligencer*, reading:

> There being many of Lieut-General Cromwell's regiment who were gone into the country to visit their friends, and could not have timely notice of his going west, by reason of the unexpectedness of that command: Its desired by the colonel that all such will repair to a rendezvous near Barking in Surrey, where they shall have further order by the officers left for that purpose.[122]

At this time, the soldiers of the armies commanded by Essex, Manchester and Waller, were in winter quarters. This usually meant they were on quarter, or billeted upon a family or garrison, but many returned home on licence from their officers on leave. Before the New Model could be moulded therefore, the soldiers had to be recalled to their colours and fresh muster rolls made. After four months in winter quarters, natural wastage through disease and disability would have reduced numbers fit for service, for this reason it was vital that a census through the muster was established without delay. On 20 February the Lords confirmed a general muster of all regiments, the soldiers being ordered to return to their colours upon pain of death, all who refused to do so to be arrested by the recognised authority in the town in question.

Likewise that same day an Ordinance was introduced to impress men as Gunners, Soldiers and Chirgions, an act that would be increasingly necessary in the coming weeks.

The 11 March saw the progress of the raising of the Army reach an important stage, for it was ordained,

> That Sir *Thomas Fairefax*, for the more speedy raising, perfecting and compleating of the Army whereof he is appointed Commander in Chief, shall have power to take into the same all such Lieutenants, Serjeants, and other Under Officers and Soldiers, as he shall think fit, as well of Horse as Foot, now or late in the severall Armies under the immediate Command of the Lord General the Earl of Essex, the Earl of *Manchester*, and Sir *William Waller*, and likewise all Gunners, Gentlemen of the Ordnance, and other Officers and Soldiers, belonging to the several Trains of Artillery in the said several Armies.

That the Lieutenants and NCOs were to be drafted before the lists of commissioned officers had been fully passed, is an indication that Fairfax was hard at work assembling the army. All military men will profess to the importance of the sergeants within regiments, for it is at this level that the daily administration of companies and troops begins. Once the NCOs were readmitted under the new commander, Fairfax had the root stock by with his army could grow and flower. Gunners too were an important element of the army, their position being semi autonomous from the general framework. The skill of the gunner was such that it was valued above that of an ordinary soldier, the chief of their trade often having extensive experience in foreign wars. In 1642 both the King and Parliament had offered considerable prestige and wages to experienced gunners, yet by 1645 the honour associated with gunnery had diminished none despite two and a half years of training.

On 17 March, the Lord Grey of Wark, who had once commanded the Eastern Association but who had been superseded by Manchester and who was now Speaker to the House of Peers, read a report to the house, outlining the position concerning the new army. He reported,

> That the List of Officers nominated by Sir *Tho. Fairfax*, and approved by the House of Commons, being sent up to their Lordships, their Lordships were pleased to disapprove of many of them; whereupon, at another Conference, the House of Commons gave their Lordships Reasons why they adhered to their Approbation; but no Answer is given to those Reasons, nor Notice taken of them. This must make the Difference endless, and the Ordinance ineffectual, which is acknowledged to be the chief visible Strength upon which we are to rely for the Safety of the Parliament and Kingdom. That Delay giveth Time to the King to recruit,

and Opportunity to break into the associated Counties; and these Officers which are mentioned in the List have Commissions, which they received from the Lord General, the Earl of *Manchester*, and Sir *William Waller*, and have been employed under them, of whom there was never yet heard any Exception against any One of them. The House of Commons agrees to their Lordships Second Reason; and from thence press, that, unless so great Reasons were to be urged against any of them as might endanger the whole Business, they conceive it requisite to concur with them, they having been hitherto employed without an Exception, and having been nominated by Sir *Thomas Fairfaxe*; and, upon those Reasons approved by them, they must not recede from their Approbation without any Reasons at all. Therefore the House of Commons again desire the Conference, to satisfy their Lordships, and all the World, that they are very desirous to use all Ways and Means to induce their Lordships to concur with the Commons of *England*, for the saving the Kingdom.'[123]

It was perfectly correct that prior to 1645 no Officer had been refused service by Parliament once commissioned by either Essex or the Association Generals, the point so tersely argued by Cromwell against Crawford's prosecution of Packer a year earlier. In no small degree, the limitation of service rested within the individual conscience, not the state.

This question then carried to the vote in the Lords as to whether the lists could pass. The wording used by Lord Grey was absolutely vital, because worded as it was, the vote would be binding to their lordships under the constitutional standing orders of their house. He asked, 'whether this House shall pass this List, as it was presented by Sir Thomas Fairfax to the House of Commons, and brought up by the House of Commons to this House.'[124] The vote being taken the result was tied, but much to the displeasure of the Manchester group, Lord Saye brought forth the proxy vote of the old Earl of Mulgrave, Fairfax's grandfather, and a supporter of the Independents. Saye had caught the members opposing the Ordinance with a trap of their own making, because they had quite obviously balanced the result to a tie, thereby having it fail on the technicality of the House which stated that tied votes were deemed to fail, it allowed this surprise proxy to carry the debate. Such a ploy could not go without question, and after further heated words it was decided that the records should be searched to see if the proxy was valid, the answer to be forthcoming the following day. A so called tacit agreement had existed in the Lords that proxy votes would not be used, but under privilege of the House tacit agreements were matters of honour rather than ruling. There is no reason to doubt that Saye had not fully researched the constitutional implications of the proxy, and knew that on 18 March it would pass to open the way for the first formative lists for the New Model Army.[125]

When during March the first properly constituted lists of Officers passed from Fairfax to discussion between the two houses, it was the plan to fully balance the character of the army. The list of Colonels passed on 18 March 1645 is of particular interest, and is included here with a list of their current command status:

Middleton	Lieutenant General to Waller
Sidney	Wounded at Marston Moor[126]
Graves	Colonel in Essex's Army
Sheffield	Colonel in Essex's Army
Vermuyden	Colonel in Manchester's Army
Whalley	Major to Cromwell/Manchester's Army
Livesey	Colonel in Waller's Army
Fleetwood	Colonel in Manchester's Army
Rossiter	Colonel serving in Lincoln
Pye	Colonel in Essex's Army

Table 1: PRESENT STATUS OFFICERS OF THE HORSE

Therefore in this first list, the Horse contained three Colonels from Essex's army, and two from Waller's, with Rossiter being all but independent yet serving in the heart of the Eastern Association. The remaining four Colonels either being loyal to Manchester, like Algernon Sidney whose severe wounds had prevented from service since June 1644, or were Cromwell loyalists like Whalley and Fleetwood.[127]

The Colonels appointed to the foot were even less controversial:

Crawford	Sergeant-Major-General to Manchester
Berkeley	Colonel in Essex's Army
Aldriche	Colonel in Essex's Army
Holborn	Sergeant-Major-General to Waller
Fortescue	Colonel in Essex's Army
Ingoldsby	Serving in Essex's Army
Montague	Colonel in Manchester's Army
Pickering	Colonel in Manchester's Army
Weldon	Colonel in Waller's Army
Rainborowe	Colonel in Manchester's Army

Table 2: PRESENT STATUS OFFICERS OF FOOT

Cromwell's Soldiers

From this it is interesting that the two junior Sergeant-Major- Generals were to be given regiments, but would obviously not retain their appointments nor titles from their former armies. It is possible that Crawford was to be a Brigade Commander in the new army, but at this time such appointments are not terribly clear in Parliamentary forces.[128] The balance of infantry Colonels from the different armies was more or less in line with the Horse, four each coming from Essex and Manchester, while two came from Waller.[129] Of particular note however, is that a quarter of the Colonels of Foot were Scots, instilling the sense that the war itself was bi-national. This balance was required if the Lords were to pass the lists, Sidney for example being a cousin to the Earl of Manchester, balanced Whalley who was related by marriage to Cromwell.

The regimental lists for 18 March show a very loose assembly of old troop and company commanders, left very much in their original formations.[130] The Lords record the new army list as follows:

FOR THE HORSE

Sir Thomas Fairfax	Colonel Middleton
Major Desborough	Major Horton
Captain Swallow	Captain Foley
Captain Browne	Captain Gardiner
Captain Lawrence	Captain Butler
Captain Berry	Captain Perry
Colonel Sheffield[131]	Colonel Fleetwood
Major Sheffield	Major Harrison
Captain Evelyn	Captain Fincher
Captain Rainborough	Captain Lehunt
Captain Martin	Captain Coleman
Captain Robotham	Captain Selby
Colonel Rossiter	Colonel Vermuyden
Major Twistleton	Major Huntingdon
Captain Markhams	Captain Jenkins
Captain Nelthorpp	Captain Bush
Captain Bushey	Captain John Reynolds
Captain Peart	Captain Middleton
Colonel Sidney	Colonel Sir Robert Pye

General Fairfax

Major Alford
Captain Dendy
Captain Nevell
Captain Ireton
Captain Bough

Major Tomlinson
Captain Knight
Captain Ireton
Captain Margery

Colonel Whalley
Major Bethell
Captain Porter
Captain Grove
Captain Horsman
Captain Packer

Colonel Graves
Major Scroope
Major General Skippon[132]
Captain Chute
Captain Doyley[133]
Captain Fleming

Colonel Sir Michael Livesey
Major Sedascue
Captain Gibbons
Captain Hoskins
Captain Pennyfeather

Captain Barry

FOR THE FOOT

Sir Thomas Fairfax
Lt. Colonel Jackson
Major Cooke Snr
Captain Cooke Jnr
Captain Beaumont
Captain Muskett
Captain Boyce
Captain Gooday
Captain Johnston

Major General Skippon
Lt.Colonel Frances
Major Ashfield
Captain Samuel Clarke
Captain Streaton
Captain Harrison
Captain John Clarke
Captain Bower
Captain Gibbon
Captain Cobbett

Colonel Holborne
Lt. Colonel Cottesworth
Major Smith
Captain Cannon
Captain Gorges
Captain Holden
Captain Wade
Captain Gorges

Colonel Crawford
Lt. Colonel Ewre
Major Saunders
Captain Eaton
Captain Smith
Captain Ohara
Captain Harvey
Captain Disney

Captain Hill
Captain Blackmore

Colonel Berkley
Lt. Colonel Emins
Major Cowell
Captain Goffe
Captain Gregson
Captain Ramsey
Captain Jameson
Captain Leete
Captain Goddard
Captain Blagrave

Colonel Aldrishe
Lt. Colonel Floyd
Major Read
Captain Wilkes
Captain Melvin
Captain Spooner
Captain Smith
Captain Wigfall
Captain Gettings
Captain Lunds

Colonel Fortescue
Lt. Colonel Bulstrode
Major Richbell
Captain Dursey
Captain Gettings
Captain Fownes
Captain Gimmings
Captain Young
Captain Yelledge
Captain Cobbett

Colonel Rainsborough
Lt. Colonel Owen
Major Donne
Captain Horsey
Captain Westome

Captain John Boyce
Captain John Binckle

Colonel Montague
Lt. Colonel Grimes
Major Kelsey
Captain Rogers
Captain Bisser
Captain Blethen
Captain Nunney
Captain Wilkes
Captain Saunders
Captain Thomas Disney

Colonel Pickering
Lt. Colonel Hewson
Major Tubbs
Captain Axtell
Captain Husbands
Captain Jenkins
Captain Silverwood
Captain Carter
Captain Gayle
Captain Price

Colonel Ingoldsby
Lt. Colonel Farrington
Major Cromwell
Captain Duckett
Captain Ingoldsby
Captain Gibson
Captain Allen
Captain Ward
Captain Mills
Captain Bamfield

Colonel Weldon

This Regiment fully passed as it came up from the House of

Captain Barber   *Commons.*
Captain Crosse
Captain Edwards
Captain Lingwood
Captain Snelling

THE DRAGOONS
Colonel John Okey
Major Gwilliams
Captain Farmer
Captain Butler
Captain Mercer
Captain Abbotts
Captain Larken
Captain Farre
Captain Bulkham
Captain Bridge

Table 3: THE NEW MODEL ARMY

If this was the army composition on 18 March, then it was constructed no differently to any army since the Scots had entered the war, with the exception of the Self Denial of those serving in Parliament. This army was certainly no revolutionary force subservient to Independency, but much would depend on how far the Covenant would act as a constraining force on the more freethinking of those chosen to lead it. It might be argued that on 18 March the army was, for the first time, modelled upon a strict Presbyterian line. In this model it was certainly designed in a way to remove the sort of confrontation seen between Manchester, Crawford and Cromwell. There was nothing in the framework which could be criticised by Crawford and his Scottish brethren, indeed it implemented many of the restrictions he had tried to introduce into the Eastern Association.

From the list it would appear that John Middleton, a Scot, was to command the Horse. He was certainly a capable officer having for all intent replaced Hesilrige as Lieutenant General to Waller in 1644. Self-Denial had removed Cromwell and Balfour, therefore the only officer still serving was this rather plain, but able, Scot. Few could deny that Cromwell was the most successful cavalry officer in the three armies, but if he could not serve, then Middleton was no bad alternative. He had served as a pikeman in France while still in his teens, and was a professional soldier serving in Essex's army at Edgehill. By November 1642 he had raised his own regiment, and continued to serve in Essex's army until a quarrel with Colonel Hans Behre 'and other officers' caused him to transfer to Waller's command.[134] At the time of the moulding of the New Model, John Middleton was twenty-six years of age, belying the old myth of youth being lacking from senior officers in Parliament's armies.

The list for the Dragoons is of interest through its association with John Lilburne. There can be no doubt that had Lilburne felt able to take the Covenant, he would, through sheer ability have commanded this regiment, but denied an open conscience he could not as we have seen serve further.[135] At this time Dragoons were seen as mounted foot, their role was to secure an area of a field, bridge or supply additional firepower to an area where it was urgently needed. Before the New Model Army, the dragoons were never required to charge as horse, yet it is evident that Okey being an experienced Captain of Horse so inspired his companies, that they would follow him into the charge.[136]

Although the Lords could through the privilege of their House delay the introduction of the officer lists, they held less influence in the Committee of Both Kingdoms where the actual physical assembly of the army was administered. For this reason, the committee was able to continue with the levying of men, a letter of 19 March to the Committee of Essex, exemplifying this programme of manpower construction

> Whereas the Parliament have by their Ordinance of 17th February, authorised you to raise, levy, and impress such numbers of soldiers for the defence of the King, Parliament, and kingdom as shall be appointed by us the Committee of both kingdoms, we finding it expedient that the army under Sir Thomas Fairfax should be forthwith completed to the full numbers, do by these presents appoint you to cause 1,000 able and serviceable men to be levied within Essex for that employment; and for your better performance of this service we have thought fit by way of direction to recommend the particulars following to your observation:-
> That especial care be taken in the choice of able, full grown, and well clothed men meet for this employment; that the new recruits be brought from the place of their levying by their conductors unto the rendezvous at St. Albans, there to be delivered to such officers as Sir Thomas Fairfax shall appoint to receive them, by the 6th April; that care be taken in the choice of conductors and assistance afforded them to keep their men from straggling and pilfering the country as they go, or from departing from their colours; that the men so impressed be commodiously provided, as has formerly been the practice, with 1,000 red coats faced with blue: and that at the delivery of the men to the conductors their numbers and names shall be certified by indenture, a duplicate whereof to be sent to Sir Thomas Fairfax. The charge of all this service the Commons have

thought fit to leave wholly to your management and good husbandry upon accompt, and have ordered you for the present to defray the same and to reimburse yourselves out of the first monthly assessment granted by the Ordinance for levying and maintaining the forces under Fairfax's command. The speedy and effectual despatch of this service so much concerns the public safety and the welfare of this kingdom, that we rest assured of your good affections and readiness, yet we cannot but earnestly desire you to use your utmost diligence and care herein that the men may be brought on the day appointed to the rendezvous, where care will be taken for the arming and otherwise providing for them.[137]

The full list of levies to be raised per county is contained in a table subjoined to the above letter:

Name of County/Town	Number of men	Colour of Coats	Rendezvous
London, Middlesex, Westminster	2,500		
Essex	1,000		St. Albans
Suffolk	1,000		
Norfolk	1,000		Watford
Norwich	60		
Hertfordshire	400		St. Albans
Ely	150		
Cambridge	250		St. Albans
Hampshire	150		St. Albans
Kent	1,000		
Surrey	350		
Sussex	600		
TOTAL	**8,460**		

Table 4: Levies to be raised by each county

The total of 8,460 men would bring the army up to strength.[138] On 20 March the Committee dealt with a number of important matters relating to Fairfax's army, but primarily furthered the moulding with endorsing acceptance by the Commons that two of Essex's regiments of horse be paid two weeks pay to prevent them being drawn away to other armies.[139] The risk that soldiers would drift to other commanders was a real one, as many of the jealousies between the old Generals had been caused by soldiers re-enlisting in different armies upon the new year. Essex's army had suffered by re-enlistment in 1643-44 to both Waller's forces and the Eastern Association. If Fairfax was not to inherit this fundamental problem, the forces to be reduced into the New Model Army, had quite simply to be provided with pay.

With Fairfax drawing up his army, at least on paper, the Parliament had to prepare his commission, for without this official blessing from the nation he was unable to command.

NOTES

[113] John Rushworth, *Historical collections of private passages of state*, 7 vols. (1659-1701) part IV, vol. 1, p.13.

[114] Richard Baxter, *Reliquiae Baxterianae, or Mr. Richard Baxter's narrative of the most memorable passages of his life and times* (1696) part I, p.48.

[115] BL Thomason Tracts E.270 [5], *Perfect Passages*, No. 17 (12-19 February 1645) p.131.

[116] *C. J.* vol. iv, pp.42, 56.

[117] 'The parliamentary diary of Lawrence Whitacre', BL Additional MS 31,116 f.195, quoted in C. H. Firth and G. Davies, *The regimental history of Cromwell's army* (2 vols., Oxford, 1940) vol.i, p.39; *C.J.* vol. iv, p.61.

[118] 'Diary of Sir Simonds D'Ewes', BL Harleian MS 166, f.179.

[119] *C.J.* vol. iv, p.63.

[120] BL Thomason Tracts E.258[1], *Perfect Occurrences*, No. 1(6-13 December 1644).

[121] BL Thomason Tracts E.260[2], *The Perfect Diurnal of some passages in Parliament*, No. 86, (17-24 March 1645).

[122] BL Thomason Tracts E.274[9], *The Moderate Intelligencer*, No. 3 (13-20 March 1645).

[123] *L.J.* vol. vii, pp.276-277.

[124] It is unclear what position Sidney held in late 1644. His poor physical condition following Marston Moor had made him unfit for field service. It is possible that his name was included in the March list simply as a sop to the Earl of Manchester.

[125] *L.J.* vol. vii, p.277. Kishlansky *The rise of the New Model Army* (Cambridge, 1979) p.46.

[126] Where the Dutchman, Vermuyden, sits in this pecking order is uncertain.

[127] The word 'Brigadier' is conspicuously absent, despite the fact that the Royalists had had brigade commanders since Edgehill.

[128] Rushworth, *Historical collections*, part IV, vol. 1, p.13.

[129] *L.J.* vol. vii, p.277.

[130] Ibid. pp.277-278. For the final list see Chapter VII, and for an analytical comparison between the March and April lists, see Appendices.

[131] James Sheffield was to be Colonel, yet his brother would be eventually named within a month.

[132] Skippon had command of a troop of horse in Essex's army, yet never served in that guise. Whether he drew pay as a Captain of Horse is unknown. In the final assessment of the New Army it was deemed unnecessary for such 'honorary troops' to continue.

[133] Charles Doyley had commanded Essex's Life-guard in 1644-45, but on 18 March a guard was not listed for Fairfax. See Barry Denton, *The Life-guard* (Leigh-on-Sea 1989).

[134] John Adair, *Cheriton, 1644: the campaign and the battle* (Kineton, Warwicks., 1973) p.181.

[135] H. G. Tibbutt, *Colonel John Okey, 1606-62* (Bedfordshire Historical Record Society Publications, 35, Streatley, 1955) p.6. Tibbutt states that John Lilburne handed over his former companies to Okey at or near Abingdon on 30 April 1645.

[136] Okey's Dragoons later became a Regiment of Horse, dragoons being largely replaced by issuing further supplies of short Carbines to the horse. Most officers had from 1642 bought a pair of horse-pistols, and Hesilrige's had carbines as early as 1643. In 1649 Thomas Horton left in his will, a carbine with rifled barrel, although these were rare at this time.

[137] *C. S. P. Dom. 1644-45*, pp.358-9.

[138] Ibid. p.359. The list was probably annexed to the letter by Rushworth when annotating the army's papers, they are however authenticated by further evidence, see ibid. p.381, Council of Both Kingdoms to the Militia of London; ibid., pp.382-383, Council of Both Kingdoms to Committee of Surrey. Ibid. p.460 gives Hertfordshire, Hampshire and Cambridge. I have in this copy inserted Watford and St. Albans for the place of rendezvous although they are left blank in the original.

[139] Ibid. p.359.

CHAPTER VI
THE BREAKING OF THE COVENANT

With the passage of the officer list remaining an issue between the Lords and Commons, the next question concerned the commission of Sir Thomas Fairfax. If Fairfax was to carry the title of General, then before he could enter the field in arms, he had to be bound in law to his appointment. In 1642 the Earl of Essex had received the military powers invested in him in the name of 'King and Parliament', the legitimacy of his commission being in a narrow point of law which maintained that Parliament was in arms to protect the kingly office and the kingdom. The argument was settled in the theoretical constitutional assumption, that Charles the man was not necessarily synonymous with Charles the King. Even if the two estates of man and monarch inhabited the one body, the kingship of Charles belonged not to himself but to the nation.[140] By this convenient hypothesis, the Parliament had convinced its more moderate supporters that it was legal to raise armies and to commission officers without the consent of King Charles I. Yet upon entering the field, the same theory would, if taken to its fullest consideration, prevent an army from bearing arms if the King was also at the head of opposing forces. This had caused considerable turmoil and some conflict of conscience in the early stages of the war, and had been the direct instigating circumstance for a number of Lords to withdraw from London to enter the King's service or attend the Oxford court.

Following the first major battle at Edgehill, and the King's subsequent move against London; culminating in the skirmish at Brentford and the display of Parliamentary strength at Turnham Green, this position was rather extended to mean that Essex and his fellow Generals were in arms to protect the King from evil council.

The Covenant had, as we have seen in a previous chapter, given a religious gloss to this position, and all who accepted it were now fighting to preserve 'Religion, King and Covenant', a wording acceptable to Essex, Waller and Manchester, but less so to sectarians and religious Independents in the army. In Fairfax's commission this covenant formula was omitted. Instead, he was placed not under 'King and Parliament', but simply under Parliament.[141] It is difficult to ascertain fully the extent to which this was a deliberate change of emphasis by the Commons, or whether the Covenant made such a reassertion unnecessary. Tthe more hawkish members of the lower house probably knew exactly the result of rewording the Fairfax commission from that received by Essex, yet such was its political imbalance within Parliament, that their true intent could not be made visible.

On 27 March the text of commission entered into debate in the Lords, and the Ordinance was twice read, but instead of passing the wording laid down by the Commons, their lordships began a further strategy to slow and hinder the foundation of the new army. Indeed with a month before the war could reasonably be expected to renew in the guise of field operations, the Lords decided to submit this vital question to a set committee. The Lords entrusted with this task were,

Earl *Northumberland*	Earl *Stamford*
Earl *Pembroke*	Earl *Sarum* [= *Salisbury*]
Earl *Nottingham*	Earl *Denbigh*
Earl *Manchester*	
L. Viscount *Saye and Sele*	
Lord *Grey*	Lord *Willoughby*
Lord *Wharton*	

Any five to meet Tomorrow morning at nine of the Clock.[142]

Each of these Lords had in 1642 accepted commissions from the Earl of Essex. Saye although being the inspiration for any progressive or radical proposals in the Upper Chamber, and also acting as patron to Sir John Meldrum and latterly Colonel Edward Aldriche, had not undertaken actual service in the field, albeit his sons had all accepted service under Essex or in the Midland Association. With the committee so structured to favour moderation, the Lords could delay transition of the commission.

The following day, probably at the instigation of Saye, the Commons sent Mr. Strode to urge swift 'Expedition, in the Ordinance of additional Power to be given to Sir Tho. Fairfax.'[143] If the Commons was to gain advantage by a speedy acceptance of the Ordinance, then it needed to yet its ordinance through the Lords within days. Any longer delay would reduce the effectiveness of the reorganisation, and therefore delay the readiness of the army for field service well into the campaigning season. This whole question of Commons urgency versus Lords sloth, was the root cause of the need for remoulding of the military administration, yet the Lords at the end of March were still slowing the very structural changes which had brought that need into being.

Instead of passing the Commons version of the commission omitting the requisite to 'preserve the King', the Lords now ordered alterations and additions, until the powers given to Fairfax were no more, yet no less, than those awarded to Essex three years earlier.[144] On 29 March the Commons sent word that, 'the House of Commons looked upon it as an Unhappiness that such a Business as this Tenderness should be in Difference between the Two Houses.'[145] They gave their reasons for not including the defence of the King in the commission:

> As concerning the Addition concerning the King's Person and etc., the House of Commons conceive that this Addition will dishearten our Soldiers; and it will encourage the King to adventure His own Person to come in the Head of His Army into Danger, which otherwise it may be He would not; also it will make the Soldiers be Judge whether the King defends Religion or not.[146]

Yet if the commission was to 'dishearten our Soldiers' then the equal effect of the inclusion of an identical clause in the Covenant must, unless revoked, subject the Army to a like danger. The last line of this statement however is more complex, for the Commons appear to suggest that if the King was himself to charge at the head of an army, he would be in breach of the Covenant responsibility to be the Defender of the Protestant Faith. This single act would place his majesty firmly against the state religion of two kingdoms, and commit him to waging war against the religious souls of his people. It is evident that such questions were indeed dangerous to the form of religious revolution envisaged in 1644-45. The words of the Earl of Manchester uttered so basely at Newbury, still echoed for all to hear nearly six months later. That it would be a further four and a half years before this impasse was broken and then only by a Parliament forcibly purged by the very army they were currently moulding into existence. It is arguable that the whole case for bringing the King to trial was adumbrated in these months of debate upon the ordinance to commission this army.

Two further pieces of legislation connected to Fairfax's powers were discussed on 29 March. Firstly he was to hold power of commission over Forts and Garrisons, which gave Fairfax the authority to name Governors and Garrison Commanders, but equally importantly to call for the surrender of Royalist garrisons standing in arms against Parliament. Secondly the Lords ordained,

> That the Committee of the Militia of the City of London and the several Subcommittees of the Militia within the Line of Communication and the Parishes mentioned within the Weekly Bills of Mortality, shall by virtue of this present Ordinance, respectively have Power, and are hereby authorized, to put in Execution, within the said City and severall Limits aforesaid, an Ordinance bearing Date the 27th of February, 1644, for the speedy raising and impressing of Men, for the recruiting of the Forces under the Command of Sir Thomas Fairfax.[147]

Compared to the question of the Covenant, this piece of bread and butter legislation, was a small matter for the two Houses, yet it is interesting that a month after its first introduction to fill the new army, impressment was to be introduced within the City of London, giving some indication that manpower was slow to muster in the proposed Foot regiments.

The Breaking Of The Covenant

More importantly on 31 March, the Lords voted upon the wording of their changes to the commission's restraints on Fairfax against the King's person. This continuation of Lordly destruction of the will of the House of Commons seemed likely to lead to a constitutional crisis. The power of the House of Commons was considered even before 1688 superior in the preparing of legislation. For the Lords to offer a second alteration upon a single issue, was, and in no small degree, challenging that primacy.

The new clause prepared by the Lords read:

> Preserving the Safety of His Majesty's Person, in the Preservation and Defence of the true Protestant Religion, the Defence of Parliament, and the Conservation of this Realm, and the Subjects thereof in Peace, from all unlawful Violence, Oppression, and Force, however countenanced by any pretended Commission or Authority from His Majesty, or otherwise.[148]

The vote was tied despite the use of Mulgrave's proxy, which for the purpose of restoring credibility between the two houses, meant the Amendments were not rejected, yet because of the constitutional requirement for a clear overall majority to allow votes to pass, the passage from Lords to Commons was blocked. That same afternoon a second reading was made of the resolution. This time, knowing they could not win the second vote, and to use the proxy would again tie the vote thus assisting the Manchester group of peers, nine of their lordships led by Lord Saye, instead of producing the proxy, desired that their dissent be recorded, a privilege granted to them by time honoured tradition. The vote was then taken:

FOR THE ADDITIONAL CLAUSE
    L. General [Essex]
    Comes *Rutland*
    Comes *Suffolk*
    Comes *Bolingbrooke*
    Comes *Manchester*
    Comes *Stamford*
    Comes *Denbigh*

    Ds. *Willoughby*
    Ds. *Bruce*
    Ds. *Berkley*

AGAINST THE ADDITIONAL CLAUSE
    Comes. *Northumberland*
    Comes. *Kent*
    Comes. *Pembroke*
    Comes. *Nottingham*
    Comes. *Sarum*
    L. Viscount *Say and Sele*

    Ds. *Wharton*
    Ds. *North*
    Ds. *Howard*

With Lord *Grey of Warke* sitting as Speaker, the question was carried ten votes to nine. It is interesting that the old Lord General the *Earl of Essex* voted in favour of the additions, giving some political speculation that he would delay the procedure even further with a refusal to lay aside his commission. If the Earl decided to maintain his seniority within the military framework, it would take a further act of Parliament to remit his commission, and Essex would remain in overall control of the army until such time that this became possible. The Lords however, possibly realising the game they were engaged upon was both dangerous to themselves, and moreover, the security of the capital, called for a Free Conference of both houses.

The evening of 31 March saw a complex series of bargaining between the two factions. Letters were sent from the lower house outlining the mutinous situation in the divided regiments currently commanded by Essex.

> The Army is in mutiny and disorder and that they know not who to obey: and until this ordinance is passed Sir Thomas Fairfax has no power to do anything...they [the Commons] have done their parts; therefore if any inconveniency come upon the stay of it, they conceive it will not lie upon them.[149]

Earlier that year, the power of the Independents had been strengthened by the addition of the Marquis of Argyle to the Scottish commissioners in London. Clarendon recalls the stance of the Marquis at this time:

> What expectation soever there was, that the Self-denying Ordinance, after it had, upon so long deliberation, passed the House of Commons, would have been rejected and cast out by the Peers, whereby the earl of Essex would still have remained general, it did not take up so long debate there. The marquis of Arguyle was now come from Scotland, and sat with the commissioners of that kingdom, over whom he had great ascendant. He was in matters of religion, and in relation to the Church, purely Presbyterian; but in matter of state, and with reference to the war, perfectly Independent. He abhorred all thoughts of peace, and that the King should ever more have the government, towards whose person [notwithstanding the infinite obligations he had to him] he had always an inveterate malice. He had made a fast friendship with sir Harry Vane during his late being in Scotland, and they both liked each other's principles in government.[150]

Throughout the debate upon the act of Self Denial, it had been the younger Vane who had guided the Independent stance, not the military experienced Cromwell. Clarendon astutely added that once Argyle came to London, the 'Scots commissioners were less vehement in obstructing the Ordinance or the new modelling the army.' This more warlike position of Argyle would assist the hawks in the Commons, his commitment to constitutional control through Parliamentary government, exceeded his Presbyterian faith in all matters political in much the same way that Vane and Cromwell paid homage to God for their victories and life, but to Parliament for all earthly powers. The disposition of the Scottish commissioners, was such that had they opposed fully the removal of the protection of the King from Fairfax's responsibility, they could by holding the Commons to the exact wording of the Covenant, uphold the objections of the Lords. It is no mere accident therefore, that Clarendon squarely cites the influence of the marquis of Argyle in bringing to fruition the Independent faction's changes in the commission.

Although debarred from military service in the new army, Cromwell, either by design or pure coincidence, had had the good fortune to be actually on military service during this latest crisis, thereby going to preserve his credibility as an innocent at large.

The events of 1 April, (All Fools Day), were upon three hundred years reflection, quite staggering. Despite denying the passage of the Ordinance the previous day, three of the anti-New Model peers were absent. According to D'Ewes, at least one of them, the Earl of Bolingbroke, chose to be absent in order to break the deadlock without having to vote for something he distrusted. The Ordinance could now pass by two votes without using Mulgrave's proxy. At ten o'clock the Free Conference met to discuss the matters that were causing so much consternation between the two houses. Shortly before the joint meeting, the Lords spurred on by the earl of Denbigh who had previously adhered strictly to the pro Essex position, approved the clause concerning the transference of Forts and Garrisons into the power of Fairfax. Following this historic moment, the earl of Essex stood to ask leave to 'Tomorrow present and deliver up his Commission to the Houses...so there will be no obstacle for the passing of the clause concerning Forts and Garrisons'. The question was put, and was passed in the Affirmative.[151] The significance of this action and vote cannot be stressed too much, for the Lords had allowed the fundamental position of the military command to pass from their hands to the hands of a Commoner. By accepting this principle of Self Denial, the Lords had transferred the militia rights previously entrusted in them alone, into the Commons via Fairfax. At no time could a Peer reassert command without first undertaking the remittance of the whole Ordinance regulating Self Denial, their role was now to structure law to be administered by the army, not to physically demonstrate that same structural administration in the field.

It is difficult to believe that Essex did not fully understand the transference of power. Yet the royalist Clarendon, witnessing only the rumour of the time, equated the whole position in April 1645 with greater clarity and insight than is available in any Parliamentary source, recording that, after it came to the House of Peers, though thereby the earl of Essex, the earl of Manchester, and earl of Warwick, and the earl of Denbigh, [whose power and authority, that is, the power, credit, and authority of the three named, had absolutely governed and swayed that House from the beginning,] were to be dispossessed of their commands, and no Peer of England capable of any employment, either martial or civil, the Ordinance found little opposition, and the old argument that the House of Commons thought it necessary, and that it would be of mischievous consequence to dissent from the House of Commons, so far prevailed that it passed that House likewise; and there remained nothing to be done but the earl of Essex's surrender of his commission into the hands of the Parliament from whom he had received it; which was thought necessary to be done with the same formality in which he had been invested with it. Fairfax was now named and declared general, though the earl of Essex made not haste to surrender his commission; so that some men imagined that he would yet have contested it: but he was not for such enterprises, and did really believe that the Parliament would again have need of him, and his delay was only to be well advised in all the circumstances of the formality. In the end, it was agreed that, at a conference of both Houses in the Painted Chamber, he should deliver his commission; which he did. And because he had no very plausible faculty in the delivery of himself, he chose to do it in writing, which he delivered to them; in which he expressed with what affection and fidelity he had served them, and as he had often ventured his life for them, so he would willingly have lost it in their service; and since they believed that what they had more to do would be better performed by another man, he submitted to their judgment, and restored their commission to them, hoping they would find an abler servant: concluding with some expressions which made it manifest that he did not think he had been well used, or that they would be the better for the change: and so left them, and returned to his house; whither both Houses the next day went to attend him, and to return their thanks for the great service he had done the kingdom, which they acknowledged with all the encomiums and flattering attributes they could devise.[152]

It is no exaggeration by Clarendon that Essex retired a bitter man, but the transition of power from the Peers to the Commons was more than the acrimonious passing from the seat of military government of one bitter man: it was far more revolutionary than that, for it transferred the war into a new dimensional setting where commoners not Lords led the army into the field.

On 2 April 1645 both the Earl of Essex, and Earl of Manchester laid down their commissions, and the following day the Lords unanimously passed the Self Denying Ordinance. With this vote, Sir Thomas Fairfax received the actual power which was his commission. The wording stated that,

Whereas the Lords and Commons assembled in Parliament have by an Ordinance, in or about the 15th Day of *February* last past, intituled, 'An Ordinance for raising and maintaining Forces, for the Defence of the Kingdom, under the Command of Sir Thomas *Fairfax*,'
Ordained.
That there be forthwith raised and formed an Army, consisting of Six Thousand Six Hundred Horse, One Thousand Dragooners, and Fourteen Thousand Four Hundred Foot, and shall be under the immediate Command of the said Sir *Thomas Fairfax* Knight, who is thereby constituted and ordained Commander in Chief of all the said Forces. The Lords and Commons now assembled in Parliament, taking the same into their serious Consideration, have Ordained, and be it Ordained, That the said Sir *Thomas Fairfax* shall rule, govern, command, dispose, and employ, the said Army, and every Part thereof, and all Officers and others whatsoever, employed, or to be employed, in or concerning the same, in, for, or about, all Defences, Offences, Invasions, Executions, and other military and hostile Acts and Services, as Commander in Chief, and be subject to such Orders and Directions as he hath or shall receive from both Houses of Parliament, or from the Committee of both Kingdoms; and the said Army, and every Part thereof, shall conduct and lead against all and singular Enemies, Rebels, Traitors, and other like Offenders, and every of their Adherents, and with them to fight, and them to invade, resist, depress, subdue, pursue, slay, kill, and put to Execution of Death, by all Ways and Means; and to fulfil and execute all the singular other Things for the governing of the said Army; and likewise shall have Power to assign and grant Commissions to all such Commanders and Officers as shall be thought necessary and requisite for the Government and Command of the said Army; and likewise shall have Power to assign and appoint One or more Provost Marshals, for the Execution of his Commands, according to this Ordinance. And it is further ORDERED and Ordained, That the said Sir *Thomas Fairfax* shall have Power to command all Garrisons, Forts, Castles, and Towns, already fortified, or to be fortified, without a moving or displacing the Governors and Commanders; as likewise, by himself or others deputed and authorized by him, to take up and use such Carriages, Draught Horses, Boats, and other Vessels, as in his Discretion, and as often as he shall think meet, shall be needful, for the conveying and conducting of the said Army, or any Part thereof, or for bringing or carrying Ammunition, Ordnance, Artillery, Victuals, and all or any other Provisions necessary or requisite for the Army, or any Part thereof, to or from and Place or Places, according to the Intent of the Ordinance; and likewise to give Rules, Instructions, and Directions, for the governing, leading, and conducting the said Army; and shall also have Power to execute Martial Law, for the Punishment of all Tumults, Rapines, Murders, and other Crimes and Misdemeanors,

of any Person whatsoever in the said Army, according to the Course and Customs of the Wars, and the Law of the Land, and according to the Laws and Ordinances of War heretofore allowed by both Houses of Parliament; and the said Laws and Ordinances of War shall cause to be proclaimed, performed, and executed; straightly charging and requiring all the Officers and Soldiers of the said Army to be obedient to him the said Sir *Thomas Fairfax*; as likewise all Lieutenants of Counties, Sheriffs, Deputy Lieutenants, Officers of the Ordnance, Justices of the Peace, Mayors, Bailiffs, and other His Majesty's Officers and Subjects whatsoever, in their respective Counties and Places, to be aiding and assisting to him the said Sir *Thomas Fairfax*, in the Execution of the said Office of Commander in Chief of the said Army, for the Ends and Purposes, and in the Manner aforesaid.

And do likewise Ordain and Declare, That the said Sir *Thomas Fairfax*, the Commanders and Officers of the said Army, and all His Majesty's Officers and Subjects whatsoever, in the Execution of the Premises, shall be saved harmless and defended by the Power and Authority of both Houses of Parliament.[153]

So ended the three-year commission of the Earl of Essex, his powers over the Parliamentary armies passing with these words to Sir Thomas Fairfax. Nowhere in the Ordinance were the words 'For the King and Parliament' to be found. Indeed the final paragraph firmly placed the King's 'Officers and Subjects whatsoever' under the rule defended by both Houses and Fairfax.

Yet if Fairfax was to command the Army under the auspices of this commission, where did it leave the military sections of the Solemn League and Covenant, for this clearly stated that,

> We shall with the same sincerity, reality and constancy, in our several vocations, endeavour with our estates and lives mutually to preserve the rights and privileges of the Parliaments, and the liberties of the kingdoms, and to preserve and defend the King's Majesty's person and authority, in the preservation and defence of the true religion and liberties of the kingdoms, that the world may bear witness with our consciences of our loyalty, and that we have no thoughts or intentions to diminish His Majesty's just power and greatness.[154]

In omitting the clause from the commission, (which called upon Fairfax, through his command, to preserve '*His Majesty's person*'), the New Model was separated by a fundamental difference of political theory from its brethren in arms in the Scottish Army under the Earl of Leven. The Scots, by their own oath, were committed to preserving the person of the King. The New Model Army now had no such legal restriction.[155]

Writing in retrospect the puritan chaplain Richard Baxter, grasped the significance of this period far better, and more openly, than many during his own lifetime or since. He wrote:

> But now begins the Change of the old Cause. A shrewd Book came out not long before, called *Plain English*, preparatory hereto: And when the Lord *Fairfax* should have marched with his Army, he would not (as common Fame sayeth) take his Commission, because it ran as all others before, [*for Defence of the King's Person*]: for it was intimated that this was but Hypocrisie, to profess to defend the King when they marched to fight against him; and that Bullets could not distinguish between his Person and another Man's; and thereunto this Claus must be left out, that they might be no Hypocrites. And this was the day that changed the Cause.[156]

This whole point must be seen as a direct leap towards a more radical view of the King's constitutional status, and a move in the war which gave the English forces greater autonomy from the Covenant. It is fair to maintain, that on 1 April the cracks in the alliance between the Anglo-Scottish forces were made visible, and on the 2nd the Covenant was broken.

NOTES

[140] For a useful discussion on the King and his constitutional position in the Civil War, see Thomas Hobbes, *Behemoth; or, the Long Parliament*, ed. Ferdinand Tönnies, with an introduction by Stephen Holmes (Chicago, 1990; facsimile of the London, 1889 edition).

[141] Mark Kishlansky, *The rise of the New Model Army* (Cambridge, 1979) p. 46. This section offers a different view to that expressed by Kishlansky.

[142] *L.J.* vol. vii p.288.

[143] Ibid.

[144] Ibid. p.289.

[145] Ibid. p.291.

[146] Ibid. p.292.

[147] Ibid.

[148] Ibid. p.297.

[149] Kishlansky, *The rise of the New Model Army*, p.47.

[150] Clarendon, Edward Hyde, Earl of, *The history of the rebellion and civil wars in England*, ed. W. D. Macray (6 vols., Oxford, 1888) book IX, vol. 4, p.4 .

[151] *L.J.* vol. vii, p.298.

[152] Clarendon, Edward Hyde, Earl of, *The Life of Edward, Earl of Clarendon*, p.298.

[153] *L.J.* vol. vii, pp.298-299.

[154] 'The Solemn League and Covenant', 25 September 1643, printed in *The constitutional documents of the Puritan Revolution 1625-1660*, ed. S. R. Gardiner (revised edition, Oxford, 1906) pp.267-8.

[155] It is significant that in 1646 the King surrendered to the Scots whose Covenant 'preserved' his person, rather than to the New Model. The letters reprinted by Thomas Edwards in *Gangraena*, (facsimile of the first edition of 1646; Exeter, 1977) go some considerable way to proving that the New Model was more radical than Kishlansky cares to admit.

[156] Richard Baxter, *Reliquiae Baxterianae, or Mr. Richard Baxter's narrative of the most memorable passages of his life and times* (1696) part I, p.49.

CHAPTER VII
AN ENGLISH ARMY

The passing of the Self-Denying Ordinance and the consequential retirement of the Earl of Essex left the framework of the militia for the New Model complete. No longer was Sir Thomas Fairfax governed by the Act which made Essex supreme Commander-in-Chief of the Parliamentary armies. He was empowered to lead the military forces commissioned by the Lords and Commons. The transition of power was not without its problems however, as Fairfax discovered. During the last days of March, while the balance of the debate turned steadily against the Essex faction, Fairfax found the senior officer list for his new army disintegrate before his eyes.

On 28 March, Edward Aldriche wrote to Fairfax objecting to the officer list appointed to command the companies of his proposed regiment of foot:

> I much mervayle at the nomination of the lyst of my regiment I shall humbly desire if possible my former lyst might stand with th'addition of a more able major and captains as my lyst may informe you, for having made inquiry I perceave some to be put to me of weak resolutions, so as I conceave, t'will be both dishonourable for me to engage with them, as also may prove prejuditiall to the cause and service we undertake, yet my lieutenant colonel in my absence hath given notice to all according to your commands, but cannot heare of Captain Smith. I have enclosed a lyst of my officers now standing with the number of soldiers, as also my opinion in two lysts concerning the new and old officers which appeares to me of soe great concernement [besides the love of the soldiers which must be lookt upon] that if not rightly stated I shall humbly desyre [though my resolution be to proceed] to desist rather then engage with dishonour, which of necessitie must follow.[157]

In reply, Fairfax, being in no position to allow Essex's old officers to dictate terms of service, asked Aldriche to stay with the regiment for an interim period, adding that he thought he would pass command to Walter Lloyd the lieutenant-colonel mentioned in his letter, and whom 'I hear much worth of'[158]. It is interesting to note that Walter Lloyd had served as a Captain to Lord Saye since before Edgehill, whereas Aldriche had been a prodigy of Lord Rochford, and favoured by Essex.[159] Aldriche had been a Major in the English army during the so-called Bishops War of 1640. In 1642 he became Lieutenant Colonel to Lord Rochford's regiment of foot.[160] Aldriche had missed Edgehill through Rochford's late marching, but joined the main army on 24 October on the road from Coventry. To complicate matters further, Rochford's regiment had been disbanded by 28 June 1643, whereupon

Aldriche found further service under the patronage of Lord Saye.[161] The role of patronage in Essex's army had played a significant part in 1642, and although the original commanders like Rochford and Saye had relinquished or never taken field service themselves, their interest in regiments continued through commissions and the influence of the peerage upon the management of the army. In the New Model however, the patronage of peers had been eroded by Self-Denial and the transition of military power towards the Commons. For the first time in three years, Oliver Cromwell's words on 'plain russett coated captains' was to be carried into the field, although a new form of patronage; the patronage of ability weighed against belief replaced the old world of lordly favour.

At approximately the same time, Algernon Sidney resigned through a leg wound which he claimed made him unfit for active field service.[162] He had indeed received a number of serious wounds at Marston Moor, where he fought under his cousin the Earl of Manchester. However, Sidney had seen most of his military service prior to 1644 in Ireland, where his conscience was less troubled in fighting against the King or fellow Englishmen, for he was fighting against Catholic rebels opposed to both King and to Parliament, and he had once declared his abhorrence of waging war in his native land. Whether this selective form of conscientious objection was part of Sidney's refusal to command in Fairfax's army cannot be determined fully, but it is interesting that the resignation came with the political weakening of Essex and Manchester's position against the Independent group. Writing to Fairfax, he claimed 'I have not left the army without extreame unwillingness...and only by reason of my lameness'. With Fairfax's consent nevertheless, Sidney was fit enough to become governor of the Sussex garrison at Chichester on 10 May 1645.[163]

Of more interest in the list of resignations, is the rather contrived departure of Colonels Middleton, Holborne, Barclay and Crawford, the professional Scottish officers who had supplied a stiff backbone to English armies. It is possible that they left through anger at the lack of Scottish regiments in the New Model, or the additional threat from the Marquis of Montrose in the Scottish Highlands, but another possibility is worthy of consideration.[164] In the debate upon Fairfax's commission, a great deal of play had been made of the wording of its position in regard to the King. In the previous chapter the argument whether the preservation of Charles the King was synonymous with Charles the man had reached the conclusion that it was permitted to treat the King as a legitimate target, and for the sake of the rules of engagement, his presence in the field did not restrict war. This had broken the binding oath laid down by the Covenant, which committed all who took it to consider that they, shall with the same sincerity, reality and constancy, in our several vocations, endeavour with our estates and lives mutually to preserve the rights and privileges of the Parliaments, and the liberties of the kingdoms, and to preserve and defend the King's Majesty's person and authority, in the preservation and defence of the true religion and liberties of the kingdoms, that the world may bear witness with our consciences of our loyalty, and that we have no thoughts or intentions to diminish His Majesty's just power and greatness.

With this commitment laid down in Scottish law, was it possible for Crawford and his countrymen to serve under the engagement in accepting Fairfax's commission? The troops commanded by old Lord Leven and David Leslie serving in northern England under the terms of the Covenant were all bound by the oath, and likewise the four Scottish officers had taken the Covenant while serving in English armies. By accepting a commission from Fairfax, were the Scots put in a position of breaking the Covenant, and thus acting in a matter perpetrating treason in their own country? In December, after all, the Earl of Essex had attempted to bring an action against Cromwell for failing to prosecute the taking of the Covenant and being an 'incendiary' in his military service. Crawford had arrested Lieutenant Packer for similar offences. By March 1645, it was clear that the new army was being moulded in a far looser form than even the sectarian splits in the Eastern Association had shown. It was emerging from its chrysalis with more of Oliver Cromwell's form than the presbyterian army favoured by the Scots. It is therefore essential to consider that the framework or constitution of the new army made service impossible for Crawford and his Scottish brethren following the 3 April votes.

The New Model therefore was an English Army, its basis being fundamentally different from the armies of 1644. It is important at this time to highlight the differences between the English army now under construction, and the Scottish army serving in England. It is also necessary to state that at no time since the Scots had entered England, had they faced a royalist army commanded by the King. They had thus far only waged war against those 'evil counsellors' who were keeping him from his people. Obviously a war fought against the Marquis of Newcastle, Prince Rupert, Sir Marmaduke Langdale and latterly the Midland garrisons, was not a war against the King's person, or at least this was the theory. Although still a year in the future, a letter to the presbyterian minister, Thomas Edwards, emphasises the differences between the feelings within the two armies:

> A Gentleman of good understanding told me lately before other company, that he meeting with a Captain of horse belonging to Cromwells Regiment [as tis commonly calld] with whom he rid some miles, and dined also at the Sheafe in Daintry the last of August, this Captain told him, that the

Parliament and Scots were agreed [that was the news] of delivering up the King to Colonel Poyntz, who was with his forces to convey him as far as Trent, or thereabouts, and then Cromwells Regiment was to bring him up to London to see him safe conveyed to the Parliament, and if he would not sign the Propositions, then he said the Parliament would decoll him, and thus they will decoll him, acting with his hand putting it to his own neck, in a way of cutting off; and this Captain added further, that he thought it would never be well with this kingdom till the King was served so. This man in his discourse speaking of Presbyterian Government, said it was a hundred times worse than Episcopal.[165]

This statement was made some fifteen months after Fairfax had undertaken his commission, yet its view was not so far removed from that taken by Lieutenant Packer in 1644, which had so alienated Lawrence Crawford. The new army was not restricted, nor handicapped, by limitations placed against it. The soldiers of the new army were therefore, free to talk of military options not open to their Scottish compatriots.

Returning to the physical changes that were forced upon Fairfax by the loss of the six Colonels, it was imperative to fill these men's boots before further commissions could be finalised and the regiments made ready for the field. With the weather improving and the first signs of Spring appearing in the fields of England, the royalists would soon be able to pull their artillery and Baggage Trayne along the muddy tracks and drovers routes, thus threatening to steal 'the march' on the still fragmented Parliamentary army. The changes were therefore made as follows:

FORMER COLONEL	NEW COLONEL
Edward Aldriche	Lieutenant Colonel Walter Lloyd
Algernon Sidney	Nathaniel Rich
Lawrence Crawford	Robert Hammond
James Holborne	Sir Hardress Waller
Harry Barclay	Edward Harley
John Middleton	John Butler

Table 5:

An English Army

Of these Lloyd was from Essex's army, Rich and Hammond having served under Essex and Manchester, while Harley and Butler were from Waller's, both having for a time served with Hesilrige. Hammond had in 1643 commanded the Life-Guard of the Earl of Essex, and was therefore perfectly acceptable to all factions of Parliament.

On 3 April, Sir Thomas Fairfax had left London to organise Army Headquarters at Windsor, where he ordered his General Rendezvous, 'having caused Proclamation to be made throughout London, for all Officers and Soldiers under his Command to give their Attendance there on Monday the 7th.'[166] The rendezvous was legal since the resignation of Essex, yet loyalty to 'Old Robin', as his troops called him, was such that many of his foot refused to leave their winter quarters at Reading. The sticking point in constructing a New Model from Essex's foot, came not from the patronymic empathy between Essex's troops and their old general alone, some of the problems came from disquiet from individuals. At the time of engaging in the new army, personal position within old regiments was not taken into consideration. Moreover, officers and NCOs in Essex's army, could, through amalgamation with Eastern Association regiments, find the three-year differential between their seniority being eroded. It was from this viewpoint, not entirely a case of disloyalty to the cause of Parliament, but a simple trade dispute between soldiers and their employer, with Fairfax being the negotiating management.

Quartered at Reading were five regiments of Essex's foot, plus five companies of Lord Robartes, at full strength a little over five and a half thousand men, but following Lostwithiel and the disease which had ravaged the army during 1644, such figures would be rather high. By 5 April it was plain that the Reading troops would not march: in military terms they were in mutiny. Yet none had freely engaged with Fairfax, and this single point so complicated the issue that coercion, not threat, was the way forward.

There was no way that Sir Thomas Fairfax could appeal to the military loyalty of these men, but Philip Skippon who had suffered, marched and fought at their side could. Skippon had been appointed Sergeant-Major-General to the new army, and was responsible for the administration of the foot. It is worth noting that Skippon had been given this command in preference to Crawford who held the appointment in the Eastern Association, or James Kerr and Henry Holborne who had intermittently commanded under Waller. Skippon was a soldier's soldier, no religious zealot, but also no Scot like Crawford, Kerr, or Holborne, and his personality was such that soldiers trusted and would follow him. On Saturday 6 April, Skippon rode into Reading, and called the mutinous regiments into the field. Setting himself at the head of the regiment previously Essex's, he addressed the soldiers not as a master, but as a friend:

Gentlemen and Fellow-Soldiers all,

> I am now to acquaint you with the Commands of Parliament, to which in Conscience to God, and Love to our Country, we are bound to give all cheerful and ready Obediance. There is a necessity lyes upon us [since three Armies are to be reduced into one] that some Commanders and Officers must go out of their Imployments wherein they now are; it is not out of any Personal disrespect to any of you that shall now go off, therefore I hope you will behave your selves accordingly.
> I have received order from the House of Commons to take notice what the Comportment shall be of those who must now go off, and to certifie unto them. I know you will behave your selves like Men of Honour and Honesty [as indeed you are] and that I shall have no cause, but to make a good Report of you, which shall do according to your Carriage in this Reducement, both to the *Speaker* of the Honourable House of Commons, and the Committee of both Kingdoms, and that particular Committee which are appointed to take you into Consideration, and to take in your Accompts, and to pay you part of your Arrears at present, and for the rest, you are to have a *Debentur* upon the Publick Faith of the Kingdom. But if there shall be any among you, who out of any Personal respect, or private discontent shall make any disturbance in this reducement, so as to obstruct or hinder a work of so publick Concernment, I shall not fail to give him his true Character to the State, without Personal respect to any. And therefore let no Man deceive himself, for although he may perhaps occasion some trouble in the present Business, yet in the issue, the greatest mischief will fall upon him himself. But I hope I shall find none such here.
> There is [at present] a Fortnight's Pay to be paid to such Officers as shall go off, which they shall presently have upon their Muster. And as for the Officers and Soldiers that are continued, [as soon as they shall be Mustered and Listed] they also shall have a Fortnight's Pay; and there are Clothes come down for the Soldiers already in part, and I assure you, that upon my Life and Honour there are more providing, and you shall be all furnished with Coats, Breeches, Stockings, Shoes, and such Arms as you shall stand in need of, with all good usage and constant Pay. And that as I have been with you hitherto, so upon all occasion of Service to my God and Country, I shall by the help of God, be willing to live and die with you.[167]

John Rushworth noted at this time that,

> This Speech and the great Esteem the Officers and Soldiers in general had for *Skippon*, had such Influence, that all freely submitted, and those five Regiments, and five Companies, were reduced into three Regiments; which work being happily effected on that Regiment which was the late General's, and so like to prove most retractory, especially for that they were Quartered in a Frontier Garrison so near the King's Quarters, which

made the matter more hazardous; they, I say, being so quietly reduced, set an Example to the rest in other Places, so that in a short time the whole Army was brought into the New designed Model.[168]

Essex's soldiers was to be the core of the new army, supplemented by men from the Eastern Association, plus a smaller number from Waller's army. On 7 April, the Committee of Both Kingdoms ordered Crawford to march his old forces to such a place that they could be reduced. He was also 'to take especial care that the soldiers be as little burdensome as may be to the country.'[169]

Skippon's reduction of Essex's army was a complete success, and on 18 April the House of Lords finally passed a list of officers. Few changes occurred between this date and the marching, but a further list from Joshua Sprigg perhaps represents the army in its final form, with only one or two notable exception. Sprigg produced his list after the battle of Naseby, and this lists Oliver Cromwell as Lieutenant-General of Horse, a post left vacant by the Lords in April. With the resignation of Middleton earlier in the month, the next most experienced cavalry officer was the Dutchman, Bartholomew Vermuyden, yet he was not named, and the army would march without a senior officer to command the horse.[170] The officer list for the New Model was entitled 'A LIST of the Names of the OFFICERS in chiefe of Foot and Horse, the Train of ARTILLERY, and other Officers, under the command of His *Excellency* Sir Thomas Fairfax; As Colonels, Lieutenant-Colonels, Majors and Captains and c.,'[171] and was constructed as follows:

GENERAL OFFICERS
His Excellency Sir Thomas *Fairfax* General
Major-General *Skippon*, Major General to the whole Army
Lieutenant-General *Cromwell*, Lieutenant General of Horse
Lieutenant General *Hamond*, Lieutenant General of the Ordnance
Commissary-General *Ireton* Commissary-General of the Horse[172]

THE TREASURERS AT WARRE, VIZ.
ALDERMEN:
Sir John *Wollaston* Knight
Thomas *Adams* Esquire
John *Warner* Esquire
Thomas *Andrewes* Esquire
George *Wytham* Esquire

ESQUIRES:
Francis *Allien*
Abraham *Chamberlan*

John *Dethyck*
Captain *Blackwell*, Deputy-Treasurer at Warres[173]
Commissary-General *Stane*, Commissary-General of the Musters
Major *Watson* Scoutmaster. General to the Army
Quartermaster General *Spencer*, Quartermaster General of Foot
Quartermaster General *Fincher*, Quartermaster General of Horse

COMMISSIONERS OF PARLIAMENT RESIDING IN THE ARMY:
Colonel *Pindar*
Harcourt *Laighton*
Thomas *Herbert*
Captain *Potter*

ADJUTANTS GENERAL OF HORSE:
Captain *Flemming*
Captain *Evelyn*

Lieutenant Colonel *Gray*, Adjutant Gen of the Foot
Captaine *Deane*, comptroller of the Ordnance
John *Mils* Esquire, Judge Advocate
John *Rushworth* Esquire, Secretary to the General and Council of Warre
Master *Boles*, Chaplaine to the Army
Commissary *Orpin*, Commissary General of Victuals
Captain *Cooke*, Commissary-General of Horse-Provisions
Master *Richardson*, Waggon-Master General.

PHYSICIANS TO THE ARMY:
Doctor *Payne*
Doctor *Strawbill*

Master *Web*, Apothecary to the Army
Master *Winter*, Chirurgion to the Generals own Person

Captaine *Wykes*, Marshal-General of Foot
Captain Richard *Lawrence*, Marshal-General of Horse

Mr. Francis *Child*, Markmaster-General of the Horse.
Master Robert *Wolsey*, Assistant to the Quartermaster-General of Foot

DEPUTIES TO THE COMMISSARY-GENERAL OF MUSTERS:
M. James *Standish*

M. Richard *Gerard*

CLERKS TO THE SECRETARY:
   M. William *Clarke*
   M. Thomas *Wragge*

MESSENGERS TO THE ARMY:
   M. Richard *Chadwel*
   M. Constantine *Heath*

FOR THE FOOT

REGIMENT APRIL 1645	PREVIOUS REGIMENT
Sir Thomas *Fairfax*, Colonel His owne Company commanded by Captain *Fortescue*	Northern Association
Lieutenant-Colonel Thomas *Jackson*	
Major *Cooke* Snr.	
Captain Samuel *Gooday*	Manchester's Rgt/M.A[a]
Captain Vincent *Boyce*	Manchester's Rgt/M.A
Captain Fulke *Musket*	
Captain *Maneste*	Manchester's Rgt/M.A
Captain Thomas *Highfield*	
Captain William *Bland*	
Captain *Lea*	

   Major-General Philip *Skippon*   Skippon's Rgt/E.A.[b]
   Lieutenant Colonel John *Frances*   Skippon's Rgt/E.A.
   Major Ashfield   Skippon's Rgt/E.A.
   Captain Samuel *Clarke*
   Captain *Streater*
   Captain James *Harrison*   Skippon's Rgt/E.A
   Captain John *Clarke*
   Captain *Bowen*.
   Captain *Gibbon*.
   Captain *Cobbet*

[a] M.A. denotes the Earl of Manchester's Army
[b] E. A. denotes the Earl of Essex's Army

Sir Hardress *Waller*, Colonel	Sir H. Waller's Rgt/W.A.[c]
Lieutenant-Colonel Ralph *Cottesworth*	Holborne's Rgt/W.A
Major Thomas *Smith*	Holborne's Rgt/W.A
Captain *Howard*	
Captain John *Wade*	Holborne's Rgt/W.A
Captain Richard *Hill*	Holborne's Rgt/W.A
Captain *Gorges*[d]	
Captain John *Clarke*	
Captain *Thomas*	
Captain *Hodden*	
Colonel Robert *Hammond*	Massey's Army
Lieutenant-Colonel Isaac *Ewre*	Crawford's Rgt/M.A
Major Robert *Sanders*	Crawford's Rgt/M.A
Captain *Disney*	Hobart's Rgt/M.A
Captain *Ohara*	Crawford's Rgt/M.A
Captain *Smith*	Crawford's Rgt/M.A
Captain John *Boyce*	
Captain John *Puckle*	Hobart's Rgt/M.A
Captain *Stratton*	
Captain *Rolfe*	
Colonel Edward *Harley*	Massey's Army
Lieutenant-Colonel Thomas *Pride*	Barclay's Rgt/E.A
Major *Cowell*	Barclay's Rgt/E.A
Captain William *Goff*	Barclay's Rgt/E.A
Captain *Gregson*	Barclay's Rgt/E.A
Captain *Sampson*	
Captain *Hinder*	
Captain *Forgison*	
Captain *Mason*	
Captain *Lago*	
Colonel Edward *Montague*	Montague's Rgt/M.A
Lieutenant-Colonel Mark *Grimes*	Montague's Rgt/M.A
Major Thomas *Kelsey*	Montague's Rgt/M.A

[c] W.A. denotes the Sir William Waller's Army

[d] This regiment formally fought under Sir Arthur Hasilrige, John Birch and latterly Holborne. Hesilrige's mother was Dame Frances Gorges, and it is likely that Captain Gorges was one of the many cousins of the regiment's original Colonel.

Captain Francis *Blethen*   Montague's Rgt/M.A
Captain Lawrence *Nunney*   Montague's Rgt/M.A
Captain John *Biscoe*   Montague's Rgt/M.A
Captain Wroth *Rogers*   Montague's Rgt/M.A
Captain William *Wilks*
Captain Thomas *Disney*   Montague's Rgt/M.A
Captain Giles *Sanders*   Montague's Rgt/M.A

Colonel Walter *Lloyd*   Aldriche's Rgt/E.A
Lieutenant-Colonel *Gray*
Major Thomas *Read*
Captain *Wilks* [possibly William Weekes]   Aldriche's Rgt/E.A
Captain P. *Gettings*   Aldriche's Rgt/E.A
Captain Benjamin *Wigfal*   Davies's Rgt/E.A
Captain John *Melvin*
Captain *Spooner*
Captain Nathaniel *Short* [possibly commissioned later]

Colonel John *Pickering*   Pickering's Rgt/M.A
Lieutenant-Colonel John *Hewson*   Pickering's Rgt/M.A
Major John *Jubbs*   Pickering's Rgt/M.A
Captain Daniel *Axtel*   Pickering's Rgt/M.A
Captain Azariah *Husbands*   Pickering's Rgt/M.A
Captain John *Jenkins*   Pickering's Rgt/M.A
Captain John *Carter*   Pickering's Rgt/M.A
Captain John *Silverwood*   Pickering's Rgt/M.A
Captain Reynold *Gayle*
Captain Thomas *Price*

Colonel Richard *Fortescue*   Fortescue's Rgt/E.A
Lieutenant-Colonel Jeffrey *Richbell*   Fortescue's Rgt/E.A
Major Thomas *Jennings*   Fortescue's Rgt/E.A
Captain Edward *Gettings*   Fortescue's Rgt/E.A
Captain Humphrey *Fownes*   Fortescue's Rgt/E.A
Captain *Young*
Captain *Gollidge*
Captain *Whitton*
Captain *Bushell*

Colonel Richard *Ingoldesby*   Tyrill's Rgt/E.A
Lieutenant-Colonel Robert *Farringdon*   Tyrill's Rgt/E.A

Cromwell's Soldiers

    Major Philip *Cromwell*
    Captain Henry *Ingoldesby*   Tyrill's Rgt/E.A
    Captain *Gibson*
    Captain *Allen*
    Captain *Ward*
    Captain *Mills*
    Captain *Bamfield*
    Captain *Grimes*

FOR THE [ARTILLERY] TRAIN
    Lieutenant-General *Hammond*, Lieutenant-General of the Ordnance
    Captain *Deane*, Comptroller of the Ordnance.
    Master Hugh *Peter*, Chaplain to the Traine
    Peter Mantreau *van Dalem*, Engineer General
    Captain *Hooper*, Engineer Extraordinary
    Eval *Tereene*, chief Engineer
    Master *Lyon*, Engineer
    Mr *Tomlinson*, Engineer
    Master Francis *Furin*, Master Gunner of the Field
    Master Matthew *Martin*, Paymaster to the Traine

    Colonel Thomas *Rainsborowe*[e]   Rainsborowe's Rgt/M.A
    Lieutenant Colonel *Bowen*
    Major *Done*   Crawford's Rgt/M.A
    Captain *Crosse*   Russell's Rgt/M.A
    Captain *Edwards*   Russell's Rgt/M.A
    Captain *Drury*   Ayloffe's Rgt/M.A
    Captain Thomas *Dancer*   Ayloffe's Rgt/M.A
    Captain *Creamer*
    Captain *Sterne*   Ayloffe's Rgt/M.A

    Colonel Ralph *Weldon*   Weldon's Kent Rgt/W.A.
    Lieutenant Colonel Nicholas *Kempson*   Weldon's Kent Rgt/W.A.
    Major William *Masters*   Weldon's Kent Rgt/W.A.
    Captain Christopher *Peckham*   Weldon's Kent Rgt/W.A.
    Captain James *Fenton*
    Captain John *Franklin*

[e] In the last week of March, Thomas Rainsborough himself was ordered to remain in Grantham and Lincolnshire in defence of the Association.

Captain Francis *Dorman*
Captain Jeremy *Tolhust*
Captain *Munday*
Captain *Kaine*

Master *Phips*, Commissary of Ammunition
Mr. Thomas *Robinson*, Commissary of the Draught-Horse

FIRELOCKS
Captain Lieutenant *Desbrowe* [Desborough]
Captain Lieutenant *Brent*
Captain *Cheese*, Captain of Pioneers

FOR THE HORSE
Sir Thomas *Fairfax*, General	Northern Association
His Troop commanded	
by Captain John *Gladman*	Cromwell's Rgt/M.A.
Major John *Desborowe*	Cromwell's Rgt/M.A.
Captain Adam *Lawrence*	Cromwell's Rgt/M.A.
Captain *Browne*	Cromwell's Rgt/M.A.
Captain William *Packer*	Cromwell's Rgt/M.A.
Captain James *Berry*	Cromwell's Rgt/M.A.
Colonel John *Butler*	Waller's Rgt/W.A.
Major Thomas *Horton*	Hesilrige's Rgt/W.A.
Captain Thomas *Foley*	Hesilrige's Rgt/W.A.
Captain Samuel *Gardner*	Hesilrige's Rgt/W.A.
Captain Thomas *Pennyfeather*	Hesilrige's Rgt/W.A.
Captain Walter *Parry*	Hesilrige's Rgt/W.A
Colonel Thomas *Sheffield*	James Sheffield's Rgt/E.A.
Major Richard *Fincher*	
Captain Robert *Robotham*	James Sheffield's Rgt/E.A.
Captain William *Rainsborowe*	Sir William Balfour's Rgt/E.A.
Captain *Martin*	
Captain Arthur *Evelyn*	E.A.
Colonel Charles *Fleetwood*	Fleetwood's Rgt/M.A.
Major Thomas *Harrison*	Fleetwood's Rgt/M.A.
Captain *Coleman*	
Captain *Selby*	Fleetwood's Rgt/M.A

Captain Richard *Zanchy*
Captain *Howard*

Colonel Edward *Rossiter*   Rossiter's Rgt/[f]
Major Philip *Twisleton*
Captain Anthony *Markham*
Captain John *Nelthorpe*
Captain *Peart*
Captain Henry *Markham*

Colonel Bartholmew *Vermuyden*   Vermuyden's Rgt/M.A.
Major Robert *Huntington*   Vermuyden's Rgt/M.A.
Captain John *Jenkins*   Vermuyden's Rgt/M.A.
Captain Henry *Middleton*
Captain John *Reynolds*   Plymouth Garrison
Captain *Bush*

Colonel Nathaniel *Rich*   Manchester's Rgt/M.A.
Major John *Alford*
Captain Jonas *Nevil*
Captain Thomas *Ireton*
Captain *Denby*
Captain *Bough*

Colonel Sir Robert *Pye*   Pye's Rgt/E.A.
Major Matthew *Tomlinson*   Abingdon Garrison/E.A.
Captain Ralph *Margery*   Cromwell's Rgt/M.A.
Captain Ralph *Knight*
Captain *Barry*
Captain Thomas *Rawlins*   Eastern Association

Colonel Edward *Whalley*   Cromwell's Rgt/M.A.
Major Christopher *Betell*   Cromwell's Rgt/M.A.
Captain Robert *Swallow*   Cromwell's Rgt/M.A
Captain John *Groves*   Cromwell's Rgt/M.A.
Captain Henry *Cannon*
Captain William *Evanson*   Cromwell's Rgt/M.A.

[f] Rossiter was already in Lincolnshire, see Chapter IX

Colonel Richard *Graves*   Graves/Essex's Rgt/E.A.
Major Adrian *Scroope*   Pye's Rgt/E.A.
Captain Christopher *Feming* [Adj. Gen]   Behre's Rgt/E.A.
Captain William Lord *Caulfield*[g]
Captain Nicholas *Bragge*   Essex's Rgt/E.A.
Captain Nathaniel *Barton*   [possibly Northern Association]

Colonel Henry *Ireton*   Cromwell's Rgt/M.A.
Major George *Sedascue*   Sir Michael Livsey's Rgt/W.A.
Captain William *Guilliams*   Waller's Rgt/W.A.
Captain Robert *Gibbons*   Sir Michael Livsey's Rgt/W.A.
Captain John *Hoskins*
Captain *Bury*

His Excellency's Life-Guard
Captain Charles *Doyley*   Essex's L.G./E.A

DRAGOONS
Colonel John *Okey*
Major Nicholas *Moore*   [possibly Nicholas Moore, Waller's Dgn]

Captain John *Farmer*
Captain Charles *Mercer*   Manchester's Dgn/M.A.
Captain Daniel *Abbots*   Manchester's Dgn/M.A.
Captain Ralph *Farre*
Captain Tobias *Bridges*
Captain Edward *Wogan*
Captain Harold *Skirmager*
Captain *Turpin*   [possibly William Turpin, Waller's Dgn]

With the officer list drawn up it was possible to assemble the regiments from the soldiers serving in the three armies, however the total numbers available in the Foot regiments were as follows:

The Earl of Essex's Army   3,048 men
The Earl of Manchester's Army   3,578 men
Sir William Wallers Army   548 men

---

[g] Lord Caulfield was a volunteer from an old Irish Peerage, and therefore was not governed or barred by Self-Denial.

This gives a total of 7,174 men available to fill regiments numbering a paper strength of 14,000. This meant that nearly seven thousand extra men had to be made up of 'pressed men' from the county quotas previously set. The need for additional men had been recognised since February, with the numbers in many regiments being extremely low.[174]

The Horse, being less reduced by the overall carnage of pike and shot warfare, were assembled from the original regiments without the need to press men, which would have been undesirable in the better quality horse. The officers of Fairfax's regiment of horse were all serving men from Cromwell's regiment of so-called *Ironsides*, the same regiment supplied five of the six troop commanders of Edward Whalley's regiment, a single troop in Sir Robert Pye's regiment, and a troop into Ireton's under its Colonel. The remaining Troops of Ireton's regiment were reduced from Waller's Army, and it is worthy of note that Butler's regiment was enlisted entirely from Sir Arthur Hesilrige's former *Lobster* regiment.[175] This meant the two Lieutenant-Generals from the Associations had supplied nineteen Troops of an establishment of sixty-seven.

On 7 April, Charles Fleetwood was ordered by the Committee of Both Kingdoms to rendezvous with Fairfax, while the north was to send 600 horse from Yorkshire to support Colonel Rossiter in Lincolnshire.[176] It was vital that the country was defended during the reductions for the new establishment. Fleetwood like Rossiter, Fairfax and Whalley, already had a regiment fit for service, with Vermuyden being already in the field with at least part of his regiment. The following day the Committee reported to Fairfax, that Sir William Waller's ex-Major, Duet, had 'left the service'.[177] Duet joined the King's service shortly after.

On 9 April, the Committee issued an interesting warrant. Captain Francis Vernon received this order.

> Out of the 2,000l. you are appointed to receive out of the Excise office, you are to pay to the officers and troopers of the regiments that were the Lord General [Essex's] and Sir Wm. Balfour's fourteen days pay, according to the establishment of 1644. And whereas divers troopers have lost their horses in the service, and are now on foot, you are, for this time, to pay them as horsemen.[178]

This move was to reduce the risk of further mutiny within Balfour's old command. That a significant number of troopers had lost their mounts, highlights a major problem with horse based armies. Simply to supply a horse for each trooper of the new establishment, required 6,100 mounts, each fit for military service. To this figure must be added replacements due to the rigours of the march, and a further supply of suitable draught animals. Needless to say the number of animals employed in military service in 1645, must have equalled the number of humans.

An English Army

Also on 9 April, came a vital direction to the Committee of the Militia of London:

> Upon complaint made to this committee of some abuses committed in impressing diverse of the Scottish nation and other strangers who are not inhabitants here, for service under Sir Thos. Fairfax, we have thought fit to desire you forthwith to give order to such as are employed that no strangers be impressed for the future, unless they be inhabitants and resident here, and all such as have been so impressed may be released.[179]

That Scots were not welcome in the new model is further demostrated by the fact that no Junior officers or even troopers were retained from what had been Crawford's and Holborne's regiments.[180] There is no doubt that Scots were from the passing of Fairfax's commission, debarred from service in the new English army, the Committee of Both Kingdoms from that time treating the soldiers of the two two nations separately.

On 17 April, order was given for the reduction of Waller's army into the New Model. Colonel Butler had still not joined Fairfax by the 24th, with 'some troops of Sir Arthur Hesilrige's regiment now lying in Surrey upon free quarter.'[181] Ireton had also not mustered the Kent troopers from Sir Michael Livesey's old regiment, Livesey receiving from the Committee numerous, often highly agitated, orders to march.[182]

An entry in the minutes of the Committee of Both Kingdoms, gives detail of the muster of the Trayne of Artillery, in the guise of a warrant from Fairfax to the Treasurers of War:

> Whereas part of the train now listed under my command were mustered at Reading on the 5th present, and the other part at Windsor and Colebrook [Colnbrook] at several different times, by reason whereof at their last muster on the 21st there were 14 days' pay due to the former and only six to the latter part. In regard it is most convenient for your accounts and best for the service that the time of their muster and pay for the whole train should be to the train, 747l.14s.113/4d, which is in full for 14 days' pay for the one part and 6 for the other part of the train.[183]

On 28 April all parishes were ordered to offer prayers for Fairfax and the new army who were to march.[184] A further order stating that,

> It is this day Ordered by the Commons Assembled in Parliament, That upon publick Notice given by Beat of Drum, and Sound of Trumpet, all Officers and Soldiers whatsoever under the Command of Sir Tho. Fairfax, all Excuses laid aside, and notwithstanding any Leave or

pretences whatsoever, do repair to their Colours by Wednesday the 30th Instant at Noon, at the furthest upon pain of Casheering for the Officers, and the Common Soldiers death without mercy. And that the Committee for the Militia of London, and Lines of Communication, and Deputy Lieutenants of the Counties of Middlesex, Essex, Kent and Surrey, do take care for the publishing of this Order by Beat of Drum, and Sound of Trumpet.[185]

Fairfax had chosen for his colours the blue of his family livery, each officer wearing a mid-blue sash or scarf to denote his alliance.[186]

The Parliamentarian news sheet, *Perfect Passages*, gives a brief account of the marching and denotes the adoption of red coats as the national colour:

There came intelligence that Sir *Thomas Fairfax* was advanced, and had but a little fit of his Ague. From Windsor he marched on Wednesday [with about 1000 Horse and Foot] to Reading, where he lay that night, and Thursday night his shouldiers expressing [with him] so much joy and alacrity in their advance, as was enough to raise the courage of any dull spirit, were such amongst them; and put valour into a very coward, to see how they did unanimously rejoice at their advance, and they marched with so much modesty and civility, that not a man amongst them durst offer wrong to any man, woman or child.

The manner of Sir *Thomas Fairfaxes* march is thus; [not to speak of Sir Thomas his own place in the march, for that is where he pleaseth] Major Generall Skippon marched in the van of the foot and leadeth the Generalls own Regiment, Sir Thomas Fairfaxes Colours are blew; the men are Red-coats all the whole Army, only are distinguished by their severall facings of their coates, the Fire-locks [only] some of them are tawny coates; and thus the Regiments of Foot march one after the other. Between every Regiment is drawn four pieces of Ordnance, and 4 the next, and so all through the Army. The Horse march some before, and some behind; some on the one winde, and some on the other as there is occasion, or as is most convenient, the carriages and traine is drawn in the middle of the Army; between the Regiments and the Pioneers; who march before the train, and make way as occasion serves, some of the Pioneers also going in every place with the Ordnance.[187]

Fairfax's illness was not uncommon among soldiers, and even senior Officers, the unsanitary conditions in the field and the close, often overcrowded, living in garrisons leading to numerous diseases becoming prevalent in the army. On Tuesday 29 April, Fairfax had had a relapse, and it was reported that 'that gallant gentleman is of such an active spirit, that it was almost death to him to be stayed so long from action.'[188]

It is obvious that *Perfect Passages* missed a nought from its imprint, for one thousand men were only a single regiment and quite insufficient for even a tentative march. Ten thousand on the other hand, was a little over fifty per cent of the establishment strength. With Fairfax and the New Model Army in the field, the civil war, after months of political wrangling, could regain its military prominence.

NOTES

[157] C. H. Firth and G. Davies, *The regimental history of Cromwell's army* (2 vols., Oxford, 1940) vol. 2, pp.384-5.

[158] Sir Thomas Fairfax to Colonel Ed. Aldrich, Windsor, 9 April, Bodleian, Tanner MS vol.60, No. 14, ff.33.

[159] *The manuscripts of His Grace the Duke of Portland preserved at Welbeck* Abbey, ed. F. H. B. Daniell (HMC, 3 vols., 1891-1923) vol. 1, pp.139, 145; *L. J.* vol. vi, pp.284, 289.

[160] *The Army Lists of the Roundheads and the* Cavaliers, ed. by Edward Peacock (1863) p.30.

[161] Peter Young, *Edgehill, 1642: the Campaign and the Battle* (Kineton, Warwicks., 1967) pp.249-50.

[162] BL Sloane MS 1519, f.112.

[163] *C.J.*, vol. iv, p.136; BL Sloane MS 1519 f.112. Jonathon Scott, in his excellent book *Algernon Sidney and the English Republic* (Cambridge, 1988) correctly points out that Sidney was offered the government of Chichester before, not after, his resignation from the New Model.

[164] Mark Kishlansky, *The rise of the New Model Army* (Cambridge, 1979) p.48, which rather skirts around the subject of the presbyterian Scots, much weakening this section of his book.

[165] Thomas Edwards, *Gangraena* (facsimile of first edition of 1646; Exeter, 1977) 'the third part', p.172.

[166] John Rushworth, *Historical collections of private passages of state* (7 vols., 1659-1701) part IV, vol. 1, p.16.

[167] Ibid. p.17.

[168] Ibid.

[169] *C. S. P. Dom. 1644-45*, pp.390-91. For an analytical assessment of the reduction see Appendix I

[170] Barry Denton, *The Regiment of Vermuyden and Cromwell* (Leigh-on-Sea 1990).

[171] Joshua Sprigg, *Anglia Redivivia; England's Recovery* (1647) pp.325-9, with augmentation from *L.J.*, C. H. Firth and G. Davies, *The regimental history of Cromwell's army* (2 vols., Oxford, 1940), and PRO paybooks. See also, G. Davies, *The Army of the Eastern Association*, [Account books 29 April 1644-1 March 1645] (1931).

[172] This appointment was later filled by Colonel Henry Ireton, which he received between 12-14 June. However, the space was blank in April.

[173] Filled by John Blackwell Junior, the twenty-one year old son of a Surrey grocer. By 1644 his father's influence in the city had secured the young Blackwell a Captaincy in the London militia of horse in Colonel Harvey's regiment. He was not a part of the April army, but probably joined it in mid-September.

[174] *C. S. P. Dom. 1644-45* gives numerous examples of reductions to fill under strength regiments.

[175] The Lobsters were curassiers raised in 1643 from a Troop originally commanded by Hesilrige in July 1642.

[176] *C. S. P. Dom. 1644-45*, pp.391, 394.

[177] Ibid. p.394.

[178] Ibid. p.395.

[179] Ibid. p.396.

[180] Ibid. p.391.

[181] Ibid. p.428.

[182] Ibid. pp.428, 447, 448.

[183] Ibid. p.432.

[184] Rushworth, *Historical collections*, part IV, vol. 1, p.25.

[185] From a 'Proclamation for all soldiers to repair to their colours', in ibid. pp.25-6.

[186] BL Thomason Tracts E.260[32], *Perfect Passages of each dayes proceedings in Parliament*, No. 28 (April 30-May 7 1645) p.218. In 1642 the Earl of Essex had chosen tawny for his colour and his officers wore sashes of that hue. In this context the word 'colour' should not be confused with flags carried by the Ensign. The portraits of Fairfax as General of the New Model Army show him wearing a blue sash.

[187] Ibid.

[188] Ibid. p.217.

CHAPTER VIII
THE CROMWELL FACTOR

The late winter, early spring of 1644-45, saw not only the transformation of the Parliamentary army into a new model, but also an important security action by regional forces. With the three main southern armies undergoing reduction into the new model, it was imperative that such regional forces restricted any adventure on the King's part to take seasonal advantage of the situation. By late March the small army which had been in the field under Waller and Cromwell, had quartered around Salisbury, while General Massey and his Western army held Waller's old campaign grounds along the Severn Valley.

Early in April, the alarm was given that Prince Rupert had taken the field with a reputed, if over-stated, 6,000 horse and foot, and lay about Worcester and the Welsh borders, with a plan to strike northward.[189] To counter this action, Major General Edward Massey marched from his headquarters at Gloucester with 400 horse and 500 foot to Ledbury, where on the 21 April he was quartered ready for any policing action needed to hinder Rupert's move north. However, the Prince, being a man who never awaited an invitation to engage in battle, arrived within half a mile of the small town on the morning of 22 April. Rushworth gives a brief account of the action which followed, recording that the Prince,

> came early on the *22nd* in the Morning within half a mile of *Ledbury*, before any discovery was made to *Massey*, who instantly caused his Horse to Mount, and drew up the Foot as well as he could; the Prince and the Lord *Loughborough* Charged into the Town several ways with great fury, and *Massey* fearing to be Hemm'd in, ordered his Foot to March away through the Enclosures with all the speed they could, and he himself with the Field Officers and some Horse entertain'd the Prince, and secured the Retreat as well as he could for three or four Miles, yet at one place, some of the Princes Horse forced their way and fell into the Rear of *Massey's* Foot, and took near 200 Prisoners in the whole action, amongst whom were two Captains, and Serjeant Major *Bacchus*, who died soon after of his Wounds: On the other side the Prince lost some Commanders, and many Common Soldiers, and was disappointed of his main End, which was utterly to have ruined *Massey*, and therefore not thinking it fit to advance Northward, retired back between *Worcester* and He*r*eford and vigorously endeavoured to supply his Army with large Recruits out of those Parts. But his Majesty in Person, with most of the Train, and a considerable party of Foot being then in *Oxford*, a Convoy of Horse, reputed about 2000 was ordered from *Worcester* and those

Parts, to fetch them off from *Oxford*, that being joined they might take the Field together.[190]

Edward Massey, despite his youth, was among the best of Parliament's Generals. His expertise in securing passage through the Severn Valley, with only a bare minimum of troops, had meant the main field armies were released from tight garrison duty and could take the advantage of their numbers.[191]

With the King in the field, Parliament could afford to wait no longer, and the Committee of Both Kingdoms had ordered Fairfax to move from Windsor, and despatch a body of Horse towards Worcester to intercept the Royalist convoy and prevent the rendezvous of the enemy forces. Despite the fact that Cromwell was at that very moment riding to Windsor from Salisbury to resign his commission to Fairfax under the terms of Self Denial, the Committee suggested he should undertake this vital task and extended his exemption from the act that he had pressed so hard for. The fact that Cromwell, not Vermuyden nor Whalley, was thought fit to undertake further action on behalf of the Committee of Both Kingdoms, when Parliament itself had by the Self-Denying Ordinance restricted his military service to but a few more days must cause some doubt as to the probability of him leaving the army at all. The question also remains that no Lieutenant-General of Horse had been appointed, albeit likely that Middleton would have served in that post had his Scottish birth and covenanting principles not exempted him. The presbyterian Richard Baxter considered Cromwell's continuation in the army to be rather more than coincidence, and noted that, 'the Confidents of *Cromwell* were especially Col. *Ireton*, and Major *Desborough* (his Brother-in-law) and Major *James Berry*, and Major *Harrison*, and Col. *Fleetwood*, and (as his Kinsman) Col. *Whalley*, and divers others.'[192]

It was now 24 April, just four days before Fairfax would have to march and prepare to administer a battle plan without the benefit of a Lieutenant General, something neither Essex, Manchester nor Waller had done throughout their commissions. It was unlikely that the Lords would agree to Cromwell being totally exempted from Self-Denial, as the earls of Essex and Manchester still held the balance of power in that House. Yet if the Committee of Both Kingdoms, consisting of English and Scottish MPs, declared its intention to use Cromwell's last allocated days of service in defence of the nation, then with Scottish backing the Committee held considerable power. Indeed, the Committee of Both Kingdoms was the powerhouse by which the future administration of the army would progress, not the two Parliamentary estates which had conceived and bore the fledgling force. It is interesting to note, that in early 1645 the power of Hesilrige, Vane and St.John was higher in the Committee than in the House of Commons, and the Independents' cause was easier to promote through this channel than the Commons.

Both Richard Baxter and Clarendon, although on different sides at this point, agreed in print, that Cromwell held considerable power in the army throughout 1644 and through 1645. Baxter writes plainly of an imagined or genuine plot on the part of the army to keep Cromwell in his position:

> And when *he* [Fairfax] was chosen for General, *Cromwell's* men must not be without him: so valiant a Man must not be laid by: The Self denying Vote must be thus far only dispensed with: *Cromwell* only, and no other Member of either House, must be excepted, and so he is made Lieutenant General of the Army: and as many as they could get of their Mind and Party, are put into Inferior Places, and the best of the old Officers put into the rest.[193]

Baxter's allegation of a plot cannot now be substantiated. Certainly there is no strictly contemporary evidence to confirm it. Certainly the Independents had wrested the legislative initiative even more from the Lords to the Commons, and had transferred some responsibility for the management of the war from the Houses themselves, where their control was spasmodic, to those executive committees, especially the Committee of Both Kingdoms, in which they had an effective majority. Whether this was, in April, a systematic plan to give them complete control of the army and of war effort is far less clear. And yet the fact that others testified later that that was what happened is worth pondering. For others shared Baxter's later view. Thus Clarendon was to write that Cromwell's place in history was secured by a pre-arranged stratagem:

> By this Self-denying Ordinance, together with the earl of Essex, the earl of Manchester, sir William Waller, the earl of Denbigh, major general Massey, lost their commands; as Cromwell should likewise have done, but as soon as the Ordinance was passed, and before the resignation of the earl of Essex, the party that steered had caused him to be sent with a party of horse into the west, to relieve Taunton, that he might be absent at the time when the other officers delivered their commissions; which was quickly observed; and thereupon orders were sent to require his present attendance in Parliament, and that their new general should send some other officer to attend that service; which was pretended to be done, and the very day named by which it was averred that he would be in the House.[194]

By 24 April, Cromwell had ridden hard to take up a defensive position in Oxfordshire near the Islip Bridge, his aim being to prevent the King from joining with Rupert at the proposed rendezvous. He still had with him six troops of his old Ironsides, who were to be Fairfax's own regiment once reduced into the New Model. John Rushworth gives a short account of the action:

> a party of Horse and Dragoons being by the General [Cromwell] ordered to March under his Command into *Oxfordshire* near *Islip Bridge* he met with a Brigade of the King's Horse consisting of the Queen's, the Earl of Northampton's, the Lord *Wilmot's*, and Col. *Palmer's* Regiments, who Engaged him, but in conclusion the Royalists were worsted, and in the pursuit many slain, and about 200 taken Prisoners, whereof several were Officers; also her Majesty's Standard, being a Crown in the midst, incircled with divers *Flower de Luces* wrought in Gold, with a Golden Cross on the top, was here taken. Some of these scattered Troops making their way to Sir *Tho. Coggins* House at *Blechington*, where Colonel *Windebank* (Secretary *Windebank's* son) kept a Garison for the King with about 200 Horse and Foot therein; *Cromwell* came up and faced that House with Horse and Dragoons, and Summoned the Governor with a sharp Message, (*Cromwell's* Troops crying out for the Foot to advance and fall on, as if there had been a Body of Foot in readiness, whereas in truth there was none) and an Answer being required to be instantly given, or else the utmost Severity to be expected, the Governor being surprised, and having no Intelligence of, or else doubting Relief from *Oxford* in time, submitted to a Parley, and consented to Surrender the House.[195]

Blechington's commander had been obviously tricked by a superior military mind, Windebank's garrison being part of the so-called 'Ring of Steel', which protected the royalist headquarters at Oxford. Cromwell had broken a quality brigade at Islip. At a similar fight at Cropredy Bridge the previous June, the Earl of Northampton had broken Waller's Army in much the same way, therefore when the broken royalist troopers thundered into the grounds of Windebank's garrison, they carried the story of the army which had caused their ignominious defeat. This was Oliver Cromwell's strategic strength; he could use his victory at Islip to cause a knock-on effect by bluffing the garrison into surrender. In taking Blechington a large supply of ammunition, horses and muskets also fell into Parliament's hands. For his lack of resistance, Colonel Windebank was tried by an Oxford Council of War, and executed by firing squad.

With still two days before the order for Fairfax's marching, Cromwell, after receiving intelligence of a party of the King's Horse were at large, rode from Blechington towards Witney. It was usual for both sides to send patrols between garrisons, in order to keep up a military pretence whereby roads might be controlled and movement held to be limited. On this morning of 26 April, Cromwell sent the Midland Association regiment of Colonel John Fiennes, the son of Lord Saye & Sele, in pursuit of a royalist patrol, and finding the King's men were transporting horses for their cavalry at Oxford. Fiennes made a brisk charge, capturing over a hundred new chargers, some Colours, and about forty prisoners.[196] The next day being the Sabbath, Cromwell sent his troopers to sneak up on the troops of Sir William Vaughan during their devotions, as Rushworth relates:

> the next day [27 April] understanding that a party of about 350 Foot, were marching under Command of Col. Sir *William Vaughan* (a Member of the House of Commons) towards *Radcot*-Bridge, *Cromwell* with a silent March, beat up their Quarters at *Brampton*-Bush, and took the said Sir *William* the Commander in Chief, the Lieutenant Col. *Littleton*, Serjeant Major *Lee*, Five Captains, a Doctor of Divinity that was Sir *William Vaughan's* Chaplain, and about two hundred Common Soldiers Prisoners, whom he sent with a Convoy to *Abington*.[197]

By capturing Vaughan, Cromwell had compelled the King to consider that only by drawing his forces together could he take the field in sufficient numbers to prevent his garrisons from sudden attack. The King had, by the last week of April, to draw out his main field army into the field, only by doing this could he make the Parliamentary army shadow him, and thereby remove such serious threat to his garrisons. To a great extent, Cromwell, had achieved his aims.

Parliament however had still not set their own campaign strategy. Some were for laying siege to Oxford, thus keeping the King bottled up in his south midland capital, and holding his Train of Artillery virtually prisoner. This plan had its advantages, for if the King could not march his supply line would be interrupted, his recruiting would be restricted, and his garrisons like Basing and Newark would remain without relief. Nevertheless, the West Country was still strongly for the King, and Taunton, which had held out for Parliament and was 'the only Inland Garrison the Parliament had in all the West of England', was 'now reduced to great Extremity, and without speedy Relief must inevitably be lost.' Possibly through financial consideration, with the Western Ports supplying much of the commerce which would pay for the new army, the latter view prevailed, and Fairfax was ordered to march westward, as we have seen, on 28 April.[198] On 29 April, Cromwell crossed the river to Farringdon, where Lieutenant Colonel Burges held the garrison for the King. Again Cromwell knew defeat for Burges was assured if relief did not arrive, but to take the garrison he would need Foot, and he called on Major-General Brown's Abingdon forces to supply that need. It would take a day or more to march the Abingdon foot to Farringdon, and in the meantime Cromwell prepared the psychological assault upon the garrison. He wrote to Burges in the following manner:

SIR,

> I Summon you to deliver into my Hands the House wherein you are, and your Ammunition, with all things else there, together with your Persons to be disposed of as the Parliament shall appoint: Which if you refuse to do, you are to expect the utmost Extremity of War, I rest
>
> Your Servant,
Oliver Cromwell

Burges refused, adding that he would not surrender without 'Order from His Majesty Himself'. Not wishing to wait too long at this garrison, Cromwell's stand hardened, and he sent a second Trumpet to Burges, writing:

> SIR,
> I Understand by forty or fifty poor Men whom you forced into your House, that you have many there whom you cannot Arm, and who are not serviceable to you: If these Men should perish by your means it were great Inhumanity surely. Honour and Honesty requires this, that tho you be prodigal of your own Lives, yet not to be so of theirs. If God give you into my hands, I will not spare a Man of you, if you put me to a Storm.
>
> Oliver Cromwell

Perhaps a little of Cromwell's sterner side to his nature can be seen here, if a 'Storm' cost him the lives of his troops, he would offer no pity. It was a trait to be seen in him in Ireland, but at a time when the military excesses of the Thirty Years War had seen the most extreme acts of barbarianism in Europe, the 'no quarter' orders given by Cromwell must be weighed lightly against them. In a reply to this threat, Lieutenant Colonel Burges wrote a staunch, and in a strangely touching way, quite inspiring refusal to capitulate:

> SIR,
> We have forced none into our Garison: We would have you know you are not now at *Blechington*. The Guiltless Blood that shall be spilt God will require at your hands that have caused this Unnatural War. We fear not your Storming, nor will have any more Parlies.
>
> Your Servant,
Roger Burges[199]

The tragedy of Civil War can be seen in Burges's words, they as Englishmen would give 'No Surrender' as their oath, the red banner would be raised and the defence of Farringdon House would begin. Cromwell would order 'No Quarter' to spur on his men, and the first volley of lead would fill the air. The next morning, at three o'clock, the Abingdon Foot marched, probably grumbling and certainly hungry, into the Town of Farringdon. Soon after their arrival, Cromwell drew up the foot with his Dragoons and began the assault, but without success, for they were repulsed with the loss of fourteen killed, and Captain Cannon finding himself on the wrong side of a scaling ladder, taken prisoner in the house.[200] It is likely that without the assistance of artillery, Cromwell's soldiers were unable to force a breach in Farringdon's walls, and rather than commit his men to a fruitless task, he withdrew to a strategic distance. At this point of time Cromwell received the news from the Committee of Both Kingdoms that Fairfax was in the field, and advancing towards him into the west. With only a further ten days before the forty day limit set by the ordinance of Self Denial came into effect, Cromwell needed further orders, for on 13 May he and the troops currently under his command, no longer held a commission. Although the soldiers serving under Cromwell were not part of the New Model, their military commander-in-chief was still Fairfax, in the same manner that Essex had been supreme officer before him.

Fairfax had marched on 30 April twelve miles to Reading, and then on May-day to Theale a further four miles, the following day reaching Newbury, where Cromwell joined him.[201] The outcome of this rendezvous would have long reaching consequences, as Clarendon explained:

> From this rendezvous the general sent to desire the Parliament that they would give lieutenant general Cromwell leave to stay with him for some few days for his better information, and without which he should not be able to perform what they expected from him. The request being so reasonable, and for so short a time, little opposition was made to it.[202]

However on 3 May, Intelligence reported movements by the royalist General Goring, and a design to attack Cromwell's position near Farringdon. John Rushworth once again records the facts:

> And that Night some Parties of Horse being sent out towards *Hungerford* and *Marlborough*, whereabouts General *Goring* was with a great Body of Horse and Dragoons coming out of the *West*, they took one Lieut. Col. *Hackett*, and some other Prisoners. By whose Examinations and other Intelligence, they discovered that General *Goring* had a design that Night or next Morning early to Beat up *Cromwell's* Quarters, whose Troops continued still near *Farringdon*; whereupon *Cromwell* immediately hasten'd back to his Charge, and had betimes put his Men into a posture to receive them; yet not so soon but that General *Goring* with a speedy March came upon the West of *Farringdon*, and recovered *Radcot*-Bridge. *Cromwell* sent a Party of Horse over the River to observe their Motions, whom the Earl of *Brainford's* Regiment, (under the Command of Lieut. Col. *Scroop*) fell upon, and repulsed them, but they being then reinforced with greater Numbers from *Cromwell*, forced *Scroop* to retire, as also a small Party of the Prince of Wales's Regiment beat back by Major *Bethel*. To whose Relief the Lord *Goring* brought part of his Regiment,

and of Col. *Richard Nevil's* (Commanded at that time by his Lieut. Col. *Standish*, Sir *Bernard Gascoign*, Capt. *Medcalf*, and other Officers) by whom the Parliamentary Party was forced to Retreat, and Major *Bethel* (venturing too far (as was said) in the dark) taken Prisoner, with two Colours of Horse, and some Common Soldiers; the Lord *Wentworth* in the interim stopping the Pass, and hindering *Cromwell's* sending over any more Forces, who yet (besides those before mentioned taken Prisoners) acknowledged not their loss to be above four or five kill'd: However General *Goring* kept his Advantage of the River, and Quartered his Horse as far back as *Lachlade*, whilst *Cromwell* making a Passage over *New-Bridge*, and having gain'd the same, General *Goring* either declining an Engagement, or rather desirous to prevent the Raising of the Siege at *Taunton* (whither Fairfax was then advancing) marched back with all speed after that Army into the West.[203]

This skirmish on the 4th allowed Fairfax to side step Goring's force at Andover, before arriving at Salisbury the next day. Cromwell, although not fighting a standing battle, had prevented Goring from joining with the King's Oxford army, simply by drawing him away from Fairfax. This factor of capability was the strength of Cromwell's position in any army he had fought in. Even during the bitterest time of argument between Manchester, Crawford and himself, Cromwell's position as Lieutenant General was safeguarded by his influence within the Independent group, but also by his ability. It is significant, and should not be underestimated, that it was Cromwell, not Balfour or Middleton who had been sent with Sir William Waller in the policing actions of early that year. Cromwell, it would seem in latter day reflection, was in the field through necessity rather than guileful opportunism, although his enemies would place this charge at his door.

NOTES

[189] John Rushworth, *Historical collections of private passages of state* (7 vols., 1659-1701) part IV, vol. 1, p.23.
[190] Ibid.
[191] Massey was born between 1617-19 and was therefore only in his mid-twenties when he became Major General to the Western Association's forces in 1645.
[192] Richard Baxter, *Reliquiae Baxterianae, or Mr. Richard Baxter's narrative of the most memorable passages of his life and times* (1696) part 1, p.49.
[193] Ibid. p.48.
[194] Clarendon, Edward Earl of, *The history of the rebellion and civil wars in England*, ed. W. D. Macray, (6 vols., Oxford, 1888) book IX, vol. 4, p.5.
[195] Rushworth, *Historical collections*, part IV, vol. 1, p.24.

[196] Ibid. p.25. The Queen's cornets were blue and from the description this was the Colonel's troop.
[197] Ibid.
[198] *C.J.* vol. iv, p.124.
[199] Rushworth, *Historical collections*, part IV, vol. 1, p.26.
[200] Ibid. pp.26-7.
[201] The full progress of the 'March' can be found in Appendix II.
[202] Clarendon, *The history of the rebellion*, book IX, vol. 4, p.5.
[203] Rushworth, *Historical collections*, part IV, vol. 1, p.27.

Cromwell's Soldiers

The Cromwell Factor

CHAPTER IX

THE PARTING OF THE ARMY

The events of late April, early May, so alarmed the King that his sole objective was now to join together his forces into an army capable of moving out of Oxford. With both Fairfax's army and the brigade under Cromwell in the field, any strategic move made by the Oxford army left the Royalists with the double threat of the army unable to withstand an open battle, whilst at the same time the outer ring of Oxfordshire garrisons were prey to tactical moves against them. Joshua Sprigg wrote of this dilemma:

> The losses the King sustained, whilest these Forces hovered thus up and down, and affronted them under their walls, so perplexed them at *Oxford*, that His Majesty sent for Prince *Rupert* and Prince *Maurice*, with all the Forces they had, to come to *Oxford*, the better to enable His Majesty to march out into the field: and for more security, General *Goring* was also sent to for that purpose, out of the West.[204]

As we have seen, Goring had been forced to make a circular march towards Oxford, but Fairfax's own manoeuvre around Newbury had left the West Country open to both the New Model and the Severn Valley brigade under Massey. The true situation was that whichever way the King turned, whatever strategy he deployed upon his own account, without the additional strength of Rupert and Maurice, the Parliamentary armies could hinder his advance.

It is interesting to note that during the first days of the New Model's march, Fairfax decided to demonstrate his determination in upholding military discipline. This was a good ploy by the General, for if he could not hold his men together through a disciplined approach, he would lose deserters at every town, and quickly earn the hatred of people wherever the army marched. In 1644 the army under Waller had been forced to hold military trials to restore order after serious breaches of military law, and it would seem it was again Waller's ex-soldiers who caused trouble.[205] Sprigg related the incident:

> In this first March of this *New Model*, the *General*, to lay an early foundation of good success, in the *punishment* of former *disorders* & *prevention* of future *misdemeanours*; caused a *Council of War* to be called, that morning that they marched from *Andover*, a mile from the Town, where the severall *Regiments* were drawn up, and stayed two or three houres: at which Councel severall Offenders were tryed for their lives: A *Renegade*, and four more Authors of the Mutinie in Kent, were

cast, one of which (whose lot it was) with the *Renegade* were executed upon a Tree.[206]

Both Ralph Weldon's and Sir Michael Livesey's (now Ireton's regiment) were from Kent, but it was the horse which had suffered some mutiny. Weldon's regiment of foot had been fully investigated by the Kent Committee, and Parliament had passed this regiment as fit for service. The regiment formerly Sir Michael Livesey's however, was at one time suspected of wanting to join the King, having previously mutinied following the battle of Cropredy Bridge. Both Sir William Waller and Anthony Weldon had accused Livesey of running away at Cropredy, the latter 'posted him for a mutineer and a coward'.[207] It is possible to piece together a defence for Livesey's troopers in 1644, because they were a Southern Association regiment forced to serve in the Midland shires. The fact that they were serving a hundred miles from their Kent homes had caused the troopers to question the legality upon which they were commanded. On 26 April, Livesey's men had been at Sevenoaks, and were ordered to march immediately to Fairfax as he marched. There can be no doubt that upon receiving this order, at least part of the regiment followed their Major, George Sedasky, in joining the New Model, while there was further mutiny among others. A similar question had hung over the Eastern Association forces ordered to march to Essex's relief at Lostwithiel, and this area loyalty above all things emphasises the need for a national standing army, or New Model. True the Earl of Essex's army had filled this requirement between 1642-45, yet Essex had never ventured beyond the Trent, and at times appeared almost transfixed within sixty miles of London. Indeed when this general area was broken, and in a fit of jealousy of Waller, the Lord General had marched into the West, it had led without doubt to his downfall.

Livesey's men had at first refused to join Fairfax as the new army was assembled, this leading to Henry Ireton being appointed Colonel in the later proposed lists. It would appear that soon after this, the regiment was brought under control by confirming the German soldier, Sedasky, as its Major, and transferring into the regiment as senior Captain, William Guilliams from Okey's Dragoons.[208] The executions near Salisbury were simply to demonstrate Fairfax's authority, and ensure that all soldiers understood the penalty for mutiny.[209]

On 2 May 1645 the Committee of Coventry wrote urgently to London:

> *May it please your Honours,*
> We have this Morning received Intelligence, That the two Princes *Rupert* and *Maurice* this last Night came with all their Forces to Evesham, and are directly on their March towards *Oxford*, and that they continue their March making no stay. This we thought our Duty to signify to you.[210]

The intelligence reports from Evesham were reliable despite the town's Royalist sympathies. In 1642 the town's Mayor had been Samuel Gardiner who now rode in the New Model with his son, under the command of John Butler, and the Parliamentary friends of the old Mayor continued to relay information to London throughout the war.[211]

On 5 May, the Committee of Both Kingdoms gave order to Fairfax to stop his march to the west:

> We send you a copy of the intelligence we have now received of the juncture of the Princes' forces with those of Goring now at or about Oxford. Considering the party left behind not to be sufficient, as we conceive, to oppose their motions and defend the Associations, we desire that Cols. Floyd, Fortescue, Ingoldsby, and Weldon, with their regiments of foot, and Cols. Greaves and Sheffield, with their regiments of horse, and such a train of artillery as you shall judge necessary, do with all possible expedition either by day or night march for the relief of Taunton, not delaying on their march. You will take order to put this party in such a way of command as may be best for the service. We desire you and Major-General Skippon, with the rest of the forces, upon intelligence held with Lieut.-Genl. Cromwell, to join in such place as may be most for the advantage of the service. The Committee of the West has written to Col. Massie to command in chief the forces that are to remain in the west. We yet know not whether he will, or his affairs will suffer him to accept it or not. If he does and comes thither, the party now sent is to be under his command, which we have signified to you that you may communicate the same to the officer who shall have the command of the party in chief. Both this Committee and the Committee of the West have written to the Western forces to receive and obey the orders of such officer as you shall appoint to command this party. We therefore desire you to send your orders with them to signify whose command they are to obey. If both or either of the regiments of Col. Cooke or Col. Thomson be come up to you, we desire they may march into the west, and in that case to bring back with you Col. Sheffield's regiment.[212]

The regiment of Sheffield was not required, but the question of command was now the stumbling block. Should command be given to Graves, or Weldon, for both were long serving officers. By 7 May the New Model had marched to Blandford in Dorset on its way to Taunton. The previous day the joint conference of the Houses had passed a vote ordering Fairfax to send only 'a Party of about 3000 Foot, and 1500 Horse and Dragoons to Taunton', the rest to turn about and join with Cromwell and Browne to cover the King's march from Oxford.[213] This order reached Fairfax while he was still at Blandford, and following his duty to Parliament he the next day drew out a Brigade consisting of Colonel Ralph Weldon's Regiment,

Colonel Richard Ingoldsby's Regiment, Colonel Walter Lloyd's Regiment, and Colonel Richard Fortescue's Regiment for the foot. The fact that four regiments were divided from the infantry, does not necessarily suggest that the regiments were considerably under strength to make the required 3,000 men. A more probable explanation was that four regiments were in the New Model the standard brigade. This would leave Fairfax with less than two further brigades of 4,000 men each with which to follow the King. It is interesting that three of the regiments of the Taunton brigade came originally from Essex's old army, with Weldon's being from Waller's. As senior officer Ralph Weldon was named brigade commander.

However, only one regiment of New Model horse was despatched by Fairfax, that of Colonel Richard Graves, the other regiments being supplied out of local garrisons and being Colonel Cooke's Regiment, Colonel Popham's Regiment, Colonel Fitz-James's Regiment, The Plymouth Horse.[214] The regiment still commanded by Graves had originally been Essex's own, and among those which broke out of Lostwithiel in 1644 with Lieutenant General Balfour. Essex's old troopers had been loath to serve a new master, and Fairfax might well have been astute in separating them from the main force.[215]

With the army divided to form a relief force for Taunton, it is worth at this time assessing the force remaining with Fairfax. The four foot regiments sent under Weldon reduced Skippon's battail, or infantry deployment, to the following:

    Colonel, Sir Thomas Fairfax's regiment
    Colonel Philip Skippon's regiment
    Colonel, Sir Hardress Waller's regiment
    Colonel Robert Hammond's regiment
    Colonel Edward Harley's regiment
    Colonel John Pickering's regiment

In the field this gave approximately some 6,000 men plus officers. The loss of Weldon's brigade reduced the army's frontage by at least three quarters of a mile when displayed in order.[216] The foot was further reduced by Montague still being at St. Albans awaiting levies for his regiment, and the fact that Thomas Rainsborough had not joined with his regiment from Grantham, to join Cromwell near Oxford.[217] It is also probable that Sir Hardress Waller had not taken command of Holborn's regiment at this time, and that his regiment, along with Hammond's and Pickering's had not made the rendezvous at Newbury-Wash a week earlier.

However, the horse with Fairfax had been vastly reduced. Not only had Graves been sent to Taunton, but Rossiter was still serving in Lincolnshire. Fairfax's own regiment of horse, formerly Cromwell's, were still at this time serving with their old commander, with whom they had faced Goring at Radcot Bridge. This left Fairfax 1,800 troopers below establishment strength. The regiments of horse therefore with Fairfax were:

>   Colonel John Butler's regiment
>   Colonel Thomas Sheffield's regiment
>   Colonel Charles Fleetwood's regiment
>   Colonel Nathaniel Rich's regiment
>   Colonel, Sir Robert Pye's regiment
>   Colonel Edward Whalley's regiment
>   Colonel Henry Ireton's regiment
>   Colonel Bartholomew Vermuyden's regiment

This gives an estimated strength on 8 May of just below 4,800 horse. This figure of approximately 8,000 men, plus dragoons, was vastly superior to anything fielded by Essex throughout 1644, yet was 12,000 below the establishment envisaged by Parliament, during the initial moulding of the army. The question set by these figures is, was the New Model Army conceived as a single field force, or was it always to be able to subdivide into brigades? The answer remains unanswered in the first hypothesis, yet through action and necessity, it is clear that any four regiments of foot could form a strong brigade, and the horse were of such general strength within each regiment to offer a capacity far superior to the regiments of Essex or Waller's command. It is with the reorganisation of 8 May that we can identify that the New Model was not simply an army apart, a miracle by which the historian might explain the change of fortune for Parliament's armies, but was more realistically a triumph of logistics. Although the religious brethren within the army would praise God for future victories, the Lord of Hosts was truly assisted by the mathematical superiority of the model set by Parliament.

Also on 8 May, the Life-guard which was commanded by Charles D'Oyley for the Earl of Essex, was re-commissioned under Fairfax, and sent from Buckinghamshire to join Cromwell. D'Oyley retained his command of 'the gentlemen' although he was to take his orders from Cromwell until Fairfax came back from the west.[218]

Weldon and Graves had marched from Blandford to Chard and advanced to Pitminster at a slow pace due to the narrow lanes. The soldier's shoes had deteriorated on the march from Oxfordshire to Somerset and the cold nights and hot West Country days were making the infantry's march difficult. Discipline and morale in Weldon's force however was high, the officers to set an example marched on foot with their men. The brigade continued its march and quartered at Pitminster, Pounsford and Trull, before continuing to Orchard on the 10th, where a royalist outpost was taken by surprise.[219] Weldon now ordered ten guns to be fired in salute, to warn Taunton of their approach, but the defenders not being able to tell the difference between friendly and enemy fire, and themselves so low in ammunition, that Blake was unable to respond in kind.

Under a false impression that Fairfax's whole army was upon them, a story probably amplified by the fliers from Orchard, the besieging Royalists, not yet supported by Goring's force, about four o'clock raised the siege. Seeing this Colonel Robert Blake assembled a force and sallied out to harass the rear of the retreating royalists but was too weak in strength to pursue them far. As no advance could safely be made from Orchard during daylight hours, Weldon ordered his forces to camp the best they could, some being quartered at South Petherton.[220] The following morning, being Sunday 11 May, the New Model attempted once more to march to Taunton, but found the roads blocked by trees cut down by the royalists for this purpose. In clearing these obstacles to allow the ordnance and supply train to pass, Weldon lost twelve hours and only covered four miles, although the Horse sent on before reached Taunton by nightfall.

On Monday 12 May, Weldon marched his Foot into Taunton, the inhabitants being overjoyed beyond expression. The soldiers saw a pitiful sight. Even the thatch had been stripped from the houses to feed the horses.[221]

Rushworth relates this progress of Weldon's brigade more simply:

> These marched directly on to *Taunton*, near which they arrived *May* 11th, upon whose approach the King's Forces before the Town raised their Siege, so that on *Monday* Morning *May* the 12th, Colonel *Welden* with the Officers without any opposition entred into it (where they found a sad Spectacle of a flourishing Town almost ruined by Fire, and Extremities of War, and the People nigh famished for want of Food) and having spent some time with Col. *Blake* the Governor, they gave Order for the whole Brigade to Retreat back and take up their Quarters at Chard.[222]

The raising of the siege was due as much to poor Royalist intelligence as to the strength of Weldon's forces. It was on 11 May reported to the besiegers that Fairfax was advancing upon Taunton with his whole army, and had indeed this been the case the Royalists would have been trapped between the town and the New Model's immense firepower. Rushworth gives some indication that the Royalists had been deliberately fooled into this belief by Fairfax's strategy in marching, by writing:

> and for this misintelligence there was some colour, for *Fairfax* with his whole Army did advance out of *Blandford* to *Wickampton* as if he had altogether design'd for *Taunton*, but then suddenly parting with that Commanded Brigade, wheeled about Eastward (marching through Inclosures, avoiding the *Champion way*, in regard, he was not well furnished with *Cavalry* if he should meet with *Goring's* Horse, who

were now upon their Retreat from *Oxford* towards the West) and coming to *Ringwood*, May the 9th; thence on the 11th to *Ramsey* (fourteen Miles;) on the 12th to *Alresford* (fourteen Miles more;) on the 13th to *Whitechurch* (ten miles;) on the 14th he reached *Newbery*, and there rested his Army three days.[223]

Therefore after splitting his force on the 8th, Fairfax took to the fields and back roads until the 12th when he reached Alresford, this giving him the time that Weldon required to reach Taunton. Rushworth does say that Fairfax 'was not well furnished with Cavalry', somewhat questioning further what regiments were with him. It is however, safer to assume that having received orders to follow the King, Fairfax was careful not to risk any action which would delay his progress towards Oxford, for it was certainly the prospect of facing the main Oxford field army, joined with the forces under Rupert and Maurice, which was the New Model's principal objective.

Back in Taunton, the population were in dire need of provision and could not quarter the extra soldiers who had marched to their relief.[224]

As we have seen however, when Fairfax had originally turned westward, Goring had quickly marched after him. It was only a matter of time therefore, before Weldon would find himself with Goring between his brigade and Fairfax's main force. There followed what can best be described as a strategic disaster for the Royalist cause, albeit not immediately apparent at the time. The Earl of Clarendon explains the situation:

> General Goring upon his return from the King, found Taunton relieved by a strong party of two thousand horse and three thousand foot, which unhappily arrived in the very article of reducing the town, and after their line was entered, and a third part of the town was burned. But this supply raised the siege, the besiegers drawing off without any loss; and the party that relieved them, having done their work, and left some of their foot in the town, made what haste they could to make their retreat eastward; when Goring fell so opportunely upon their quarters that he did them great mischieve, and believed that in that disorder he had so shut them up between narrow passes that they could neither retire to Taunton nor march eastward: and doubtless he had them then at a great advantage, by the opinion of all men that knew the country. But by the extreame ill disposing his parties, and for want of particular orders, (of which many men spoke with great license,) his two parties sent out several ways to fall upon the enemy about Petherton-bridge, the one commanded by Colonel Thornhill, the other by sir William Courtney, (both diligent and sober officers,) they fell foul on each other, to the loss of many of their men; both the chief officers being dangerously hurt, and one of them taken, before they knew their error; through which the enemy with no

more loss got into and about Taunton: notwithstanding which untoward accident, general Goring was, or seemed, very confident that he should speedily so distress them that the place would be sooner reduced by the relief that had been put into it, and that in few days they would be at his mercy.[225]

The same story is told by Joshua Sprigg:

About this time came news of a remarkable passage in Gen. *Goring's* army in the West; which as will afterwards appear, hath been their lot to happen among them more than once: Namely, a hot skirmish, which one party of his horse had with another party of his own horse, neer *Crookhorn*, thinking they had been ours, (For indeed a party of horse of Col. *Welden's* Brigade were then within a mile of them at *Hinton St. George*:) in which skirmish many of the Enemy were slain, both Officers and Souldiers, by one another; and that party of their horse that was routed, fled as far as *Bath*, giving a hot alarm as they went, which for the present put them in some distraction. Providence had ordained this accident, as an advantage for that party of our Horse, who otherwise might have been endangered (by the sudden advance of the Enemies forces) in their retreat from *Pederton* to *Taunton*.[226]

It was not uncommon for troops to mistake each other for the enemy, which was primarily why all New Model infantry wore the prototype redcoat that would so identify the English and later British soldier. From Clarendon's account, it is obvious that Weldon was trying to rejoin Fairfax, but was now trapped in Taunton. It is unknown how much by way of provisions the New Model had transported to Taunton, but with five thousand extra mouths to feed, Goring was obviously confident that the Parliamentary garrison could not hold out. Clarendon continues by estimating that the force with Goring was no less than 5000 foot and 4000 horse, which was not considerably under the numbers currently with Fairfax. The question remains whether these forces now under Goring would have been better used by the King.

On 14 May, the Committee of Both Kingdoms wrote directly to Fairfax ordering him to rest the army, and the next day wrote again to congratulate his forces on their success at Taunton. Yet their victory threatened to become a defeat, and within days Weldon's brigade had found themselves in Taunton as part of the defending force.[227] However, if Weldon could not stand against Goring in the field, neither could the Royalist commander now hope to penetrate the defences of the old town. Weldon's brigade consisted of both pike and shot, therefore if his forces were balanced with about fifty percent of each arm, at least 2,000 additional musketeers had been added to the defences. This extra firepower would prevent

Goring from taking the town by storm, but would now result in both Royalist and Parliamentarian forces being trapped in the guise of besieged and besieger.[228]

Goring now began to renew the siege, yet after months of jealous bickering between the officers of the Western command, he was far from being in a position in which loyalty could be expected from the soldiers under him.[229] By 24 May the King had written to Goring to march to him for better service with the field army. Clarendon tells the whole sorry tale which followed:

> In this conjuncture the King's letter came to the Lord Goring to march; to which he returned an answer by an express before he desired the Prince's directions, though he was diligent enough to procure his highness' opinion for the respite of his march. And the truth is, the assurance that he gave of his reducing those forces within very few days; the leaving all the west to the mercy of the rebels, if he went before they were reduced; the danger of their marching in his rear, and carrying as great an addition of strength to the enemy as general Goring could carry to the King, except he carried with him the forces of the several garrisons which were then joined to him; made it very counsellable to suspend a present obedience to those orders, till his majesty might receive the full and true state of his affairs in those parts, to which purpose an express was sent likewise by his highness to the King.[230]

Of course Goring was correct in saying that by leaving Taunton it would also add strength to Parliaments forces, but whereas the majority of the Royalists were cavalry, only Grave's regiment were horse from the New Model. From a tactical viewpoint, it is therefore worth consideration, that Weldon's brigade was cutting off a vital supply of cavalry from the King's main field army. This stance by Goring was further criticised by Clarendon who reported that:

> In the mean time general Goring was so far from making any advance upon Taunton, that he grew much more negligent in it than he had been; suffered provisions in great quantities to be carried into the town through the midst of his men; neglected and discouraged his own foot so much that they ran away faster than they could be sent up to him; and gave himself wholly to license, insomuch that he many times was not seen abroad in three or four days together.[231]

While Fairfax divided his army, the King slipped from Oxford for the opening of his spring offensive. John Rushworth gives a brief detail of the Royalist movements:

> His Majesty on the 15th of May came to *Hemley*, two miles beyond *Sturbridge* in *Worcestershire*, and there and at *Droitwich* rested three or

four days, whilst Prince *Rupert* took in *Hawkesly* House (a small Garison of the Parliaments, about seven or eight miles) which being Surrendered was Burnt. From *Hemley* His Majesty advanced to Wolverhampton, and from thence May the 17th to *Newport* in *Shropshire*: where by the way Sir *William Vaughan* Governor of *Shrawarden* Castle, (two miles from *Shrewsbury*) coming towards *Bridgnorth* to meet his Majesty, fell upon some Shrewsbury Horse at Wenlock and worsted them.[232]

On 15 May the New Model force still with Fairfax was weakened further by the Committee of Both Kingdoms, whose orders stated:

> We have appointed a Party of Two thousand Horse, and five hundred Dragoons to march to the *North*, for the assistance of the Scotch Army, to be made up of the Regiments of Col. *Sidney*, Col. *Vermuyden*, Sir Robert *Pye*, and Col. *Fiennes*; and in case they shall not make up 2000, then to fill that Number up with such of the Troops now under the Command of Major *Sadasky* [Sedascue] as are fullest: That these and 500 Dragoons all under the Command of Col. *Vermuyden* shall march towards the *Scotch* Army, in case the King's Army shall march Northward. The rest of the Forces that are left with Lieutenant General *Cromwell* are appointed to march back towards *Blechington* to put a Garrison into that House, and dispose the Forces in the best way that may be for the streightening of *Oxford*, which we have design'd to be Blocked up forthwith, in order to a Siege, with those Forces that are at present with you, and with Lieutenant General *Cromwell*, and with all the Recruits that are to come up, and such other Forces both of the Garisons and other Counties as we shall send thither for that purpose.
>
> We have designed this as the main, *Abington* to be preserved; with the rest of its Forces not already elsewhere appointed, unless there shall any especial Exigency require them to be otherwise imployed: We desire you therefore so to dispose of your March as may be of most advantage to the design of Blocking up and Besieging of *Oxford*. We desire you to consider what is necessary for such a work, and to advise us thereof.[233]

During the first few days of its march, Colonel Bartholomew Vermuyden's brigade manoeuvred around Shewsbury, keeping a careful watch over the King's movements. Vermuyden had been sent north to join with the Scots, and subsequently with Colonel Rossiter, but had been by circumstances forced to survey the King, before turning east where the Lincoln forces were deployed. In late May Vermuyden's brigade was at Posenhall thirteen miles southeast of Shrewsbury, moving to Birmingham has the King made his advance to Wolverhampton.[234]

The regiments of Vermuyden, Sidney and Pye numbered at full strength 1,800 men. It is interesting that Lord Say, who wrote the letter quoted above, calls Nathaniel Rich's regiment, Sidney's, this being its commission of 18 March rather than its April listing, perhaps signifying the stress that the Committee was working under. Interestingly, Say also confirms that Ireton had yet to join Fairfax, and that George Sedascue was still commanding the Kent horse as Major.

This latest division of Fairfax's force left the General with only four regiments of horse, which in itself is evidence enough that by the third week of May, the Committee of Both Kingdoms was confident in their ability to put the central part of the New Model into a static position about Oxford. It is clear that Fairfax was General in the field, but the Committee was nothing less than a prototype 'War Cabinet' administering the strategic elements of its army.

The question was now with the King having left Oxford, how quickly could Fairfax's campaign plan be put into action. Rushworth relates,

> For the Reader's better understanding the Military Transactions of this juncture he must note, that upon *Fairfax's* being recall'd, and return with the greater Party of his Army out of the West, the Houses considered whether he should be Imployed to sit down before *Oxford*, or follow the King, who seemed to bend Northwards.[235]

Parliament was already paying a Scottish Army of 21,000 Horse and Foot to guard the north, a force it is worth remembering equal in strength to the New Model. In addition there was a considerable force of Parliamentarian's in Cheshire under the command of Sir William Brereton, whose army was blocking the King's passage into North Wales by besieging Chester.

NOTES

[204] Joshua Sprigg, *Anglia rediviva; or England's recovery* (1647) p.13.
[205] The Court Martial papers of Sir William Waller's army can be found in John Adair, *Cheriton, 1644: the campaign and the battle* (Kineton, Warwicks., 1973).
[206] Sprigg, *Anglia rediviva*, p.15.
[207] Anthony Weldon, *The True declaration of Colonell Anthony Welden* (1645) p. 16, mispag. 14. At the restoration, Denzil Holles quite amazingly transferred the charges laid against Livesey to Sir Arthur Hesilrige, but at a time when the latter was long dead.
[208] C. H. Firth and G. Davies, *The regimental history of Cromwell's army* (2 vols., Oxford, 1940) vol. 1, p.116. Sedascue was a religious refugee from Czechoslovakia. He was an excellent soldier having served in 1642 as cornet to George Urry. When Urry had turned coat, Sedascue sought service with Livesey with who he became Captain, then Major. Guilliams had served under Sir Arthur Hesilrige, then in Waller's own regiment as Captain. In the 18 March lists, Guilliams was given service in the dragoons, but his experience as a cavalry officer must have swayed Fairfax in transferring him to Ireton's command.

The Parting Of The Army

[209] Ireton's regiment, despite being fully under control, continued to follow a path toward a disregard of authority, and a tendency for mutiny. They joined the crisis of 1647, and the mutiny of 1649.

[210] John Barker to the Committee of Both Kingdoms, 2 May 1645, quoted in John Rushworth, *Historical collections of private passages of state* (7 vols. (1659-1701) part IV, vol. 1, p. 28. The messenger from Coventry would have taken a day to reach London, then a further day to ride to Fairfax. It took until 6 May for the Houses to pass an order for Fairfax's own moves, therefore wasting two days in administration.

[211] A document relating to Samuel Gardiner as Mayor survives in Worcester and Hereford Record Office, covering the needs of the many wounded following Edgehill. Gardiner had volunteered to serve under Lord Brooke in 1643, he joined Hesilrige on the death of Brooke at Litchfield, before being reduced into Butler's regiment in the New Model. His son, another Samuel later served as Mayor of Evesham.

[212] *C. S. P. Dom. 1644-45*, p.460.

[213] Rushworth, *Historical collections*, part IV, vol. 1, p.28.

[214] Ibid. Colonel Nicholas Boscowen took command of the 'Plymouth Horse' vacated by John Lutteral in October 1644. Boscowen had formerly from 5 August 1644 commanded a troop of Dorset Horse forming the Garrison at Wareham. Boscowen's service was short but distinguished, he died at Durham House and was buried at Westminster Abbey on 22 September 1645. Boscowen's officers were:

    Colonel Nicholas Boscowen
    Captain Lieutenant Joseph Underwood
    Major Richard Stevens
    Captain George Walters
    Captain William Braddon
    Captain Jobst William Van Jaxehein
    Captain Onesimms Penrose
    Captain James Dewy
    Cornet Anthony Coombe

It is interesting that upon Boscowen's death the regiment on 23 September passed to Sir Francis Drake.

[215] From May 1645-May 1649, the regiment of Graves and Scroope was with the main army for less than six months, which was in 1647, when it was part of the Guard for the King at Holdenby, before being sent to the West yet again.

[216] The deployment field plan shown by Joshua Sprigg at Naseby includes the regiments of Montague and Rainsborough. Taking this as Skippon's battle plan, the New Model in early May was unable to field more than four regiments of foot in its front line, with two reserves.

[217] *C. S. P. Dom. 1644-45*, p.468.

[218] Ibid. pp.468-9.

[219] Emmanuel Green, *The Siege and Defence of Taunton* (Somersetshire Archaeological and Natural History Society's Proceedings, 1879) vol. XXV, pp.33-48, quote on pp.15-16. Many thanks to Somerset County Record Office for this reference.

[220] Ibid. pp.16-7.

[221] BL Thomason Tracts E.45[6] *Perfect Occurrences of Parliament*, No.19 (3 May 1644); John Vicars, *Magnalia Dei Anglicana, or, Englands Parliamentary Chronicle* (1646) the fourth part entitled 'the burning-bush not consumed', pp.147-148.

[222] Rushworth, *Historical collections*, part IV, vol. 1, p.28.

[223] Ibid. p.29.

[224] Vicars, *Magnalia* ['the burning-bush] p.148. See also *A Great Victorie obtained against the enemy at the Raising of the Siege from before Taunton* (1645).

[225] Clarendon, Edward Earl of, *The history of the rebellion and civil wars in England*, ed. W. D. Macray (6 vols., Oxford, 1888) book IX, vol. 4, p.45.

[226] Sprigg, *Anglia rediviva*, pp.22-3.

[227] *A Narration of the Expedition to Taunton* (1645) reproduced in Appendix IV. This account suggests that Fairfax had only three Regiments of Foot with him after dividing his forces at Blandford, his own, Skippon's and Harley's.

[228] The siege of Taunton cost the New Model heavily in losses to this brigade, including Walter Lloyd, whose regiment appears to have taken particularly heavy casualties during the six weeks of service in the town. For an analytical assessment of all casualties during 1645-46 see Appendices.

[229] For a full account of the rather petty jealousies between the Royalist western commanders, see Clarendon, *History of the rebellion*, book IX, vol. 4. Although Clarendon is often regarded as biased against his contemporaries, there is no reason to doubt that his account of the squabbling within the command structure is anything but accurate.

[230] Ibid. book IX, vol. 4 p.47.

[231] Ibid.

[232] Rushworth, *Historical collections*, part IV, vol. 1, p.29.

[233] Ibid. p.32.

[234] *The letter books, 1644-45, of Sir Samuel Luke, Parliamentary General of Newport Pagnell*, ed. H. G. Tibbut (Publications of the Bedfordshire Historical Record Society 42; HMC joint publications, 4, 1963) pp.543, 551.

[235] Rushworth, *Historical collections*, part IV, vol. 1, p.29.

CHAPTER X
OXFORD AND LEICESTER

Although Sir Thomas Fairfax was commander-in-chief of land forces, it had been recognised by Parliament during the moulding of the new army that if they were to retain effective control over the national forces, a secondary institution of parliamentarian commissioners would be required to accompany him and to report back to them in matters of administration between the military and civil authority. It is true that commissioners had resided with the armies since 1642, yet during this time the senior commanders had themselves been either members of the Lords or Commons, thus alleviating the need for the Houses and their executive committees to keep a watchful eye on their forces.

On 16 May the rules governing the 'Commissioners of Parliament appointed to reside in the Army' were read in both houses, and (the Lords concurring with the Commons) the instructions were passed as follows:

> 1. That for the prevention of *False Musters*, they are to see that the Advocate of the Army administer an Oath to such Persons attending upon Musters, as shall be presented unto him for the discovery of false Musters, or accusing any Person that shall violate the Articles of War in the case of Plunder, or otherwise.
> 2. For the ease of the Country the said Commissioners are to endeavour that no Officer or Soldier be Quarter'd in any Place but by the Quartermaster, first shewing his Commission, if it be required, and by what Authority he takes up such Quarters; and giving a Ticket of the Names of every Person which he shall Quarter, expressing of what Regiment, Troop or Company the same Person so Quarter'd is; and the Number of Horses there Quarter'd, and at whose House; together with the day of the Month, and to subscribe his Name thereunto: Saving where by reason of the great Numbers of them they cannot be Incerted, and then their Numbers to be expressed in place of their Names.
> 3. That no Quarter or Provisions for Man or Horse in any Quarters be taken without Payment of ready Money; but in case of Necessity for want of Pay; which the Parliament will use all means possible to prevent. And in case and Quarter or Provision shall be taken without payment, the Captain or Quarter-Master shall by writing under their or one of their hands, certifie what Provisions have been so had, within that time, by whom, and of what Regiment, Troop and Company, from whom and the value thereof, Provided, that where the Army shall be upon their March, not staying above twenty four Hours in a place.[236]

The payment rates for each provision made in Quarter were also to be in strict accordance with sums allocated by the Parliament, emphasising the pedantic nature of the embryonic Civil Service requirement:

Commodity	Price
Hay	4d per night
Grass	3d per night
Oats	4d a peck
Peas or Beans	6d a peck
Barley and Malt	7d a peck

FOR THE DIET OF:

Trooper	8d per day
Dragooner	7d per day
Foot-Soldier	6d per day
Pioneer	6d per day
Waggoner or Carter	6d per day

Table 6: Payment rates for each provision made in Quarter.

Officers were traditionally expected to pay their own bills of Quarter, this being allowed for in their pay, and a sign of their superiority of social status over the rank and file. Likewise the gentlemen who made up Fairfax's Life-Guard Troop were, as gentlemen, expected to pay for the Quarter of themselves and their mounts, and quite naturally any servants employed by them.

It is interesting however to consider that the 600 troopers of a New Model regiment were officially allocated a maximum of 1s 3d each to feed themselves and mount per day on Quarter, or £37.10.0 maximum per regiment. The rules for Quarter also maintained that,

> Provided also, that no Inhabitant whatsoever shall be compelled to furnish any Provision but what he hath in his House of his own: And that no Officer or Soldier shall compell him to do otherwise, upon pain of Cashiering, or such other punishment as the Commander-in-Chief shall think fit.[237]

It is clear that for the first time Parliament was attempting to regulate their military power in a professional manner. Yet considering Fairfax was already in the field, these actual administrative details were overdue.

While Parliament finalised the finer details of the Commissioners' remit and sent Colonel Pindar, Captain Potter, Harcourt Leighton and Thomas Herbert to join Fairfax as the said four administrators, the General continued his march towards Oxford. On 17 May Lord Saye, on behalf of the Committee of Both Kingdoms, wrote to Fairfax, who lay that night at Blewbury after a ten-mile march. Saye informed the General thus:

> SIR,
> We wrote unto you the Fifteenth Instant, giving you notice of our intention for the present Blocking up and future Siege of *Oxford*, by the enclosed Copy of the Order of Both Houses; you see it is now their resolution, who have written to the Committees about it; we therefore desire you forthwith to send such of your Horse as you can spare, leaving sufficient to bring up your Foot, to prevent the carrying of any Provisions into that Town; we have written also to that purpose to Lieut. General *Cromwell* and Major General *Brown*. We desire this may be speedily done, and withal to send us word what Provisions you think necessary for that Siege, that there may be a speedy Provision thereof. In regard this design is now publick, we desire you to make all speed to send in some Horse to hinder the carrying in of Provision to *Oxford*, and the burning and spoiling of the Country.[238]

Fairfax had only four regiments of Horse still with him, the rest of the establishment being with Vermuyden, Cromwell, or trapped in Taunton under Graves. The regiments of horse still with Fairfax were Colonel John Butler's regiment, Colonel Thomas Sheffield's regiment, Colonel Charles Fleetwood's regiment, and Colonel Edward Whalley's regiment. These four regiments numbered no more than 2,400 officers and men, he would need at least two regiments to form a protective guard to the Foot and Trayne, plus Troops divided from the body to act as patrols. Therefore if Fairfax was to send an advance body of Horse towards Oxford it would through necessity be small, possibly no more than 800 men.

By 19 May the Foot had reached Newnham in Oxfordshire, and the next day reaching Garsington, and by the 22nd Marston on the outskirts of the City of Oxford. Here the royalist Captain Gardiner with a body of Horse and Foot charged a party of Fairfax's Horse under Adjutant-General Flemming, but was repulsed with near two hundred of the Royalist Foot taken.[239]

Sir Thomas Fairfax could now control the outer defences of Oxford, the city being effectively cut off from supply routes to the west. A breastwork was quickly thrown up by the New Model pioneers on the east side of Cherwell, with a bridge secured near Marston. Added to Fairfax's forces, were the Horse and Dragoons under Cromwell, and the forces under Brown, giving enough Horse and Foot to lay siege to the city once the heavy artillery pieces were brought up from surrounding garrisons.[240] To this end the Committee for the Army ordered on 23 May that:

> We desire you to see that the following pieces of ordnance, viz. 1 demi-cannon and 1 whole culverin at Windsor, 1 demi-cannon and 2 whole culverins at Northampton, and 1 piece called the pocket pistol at Cambridge, be put into present readiness to be brought to such a place as Sir Thomas Fairfax may appoint for the service against Oxford.[241]

The pocket pistol was one of a pair of Royalist pieces captured the previous year, and variously called 'The Queens Pocket Pistols', or Gog and Magog from the giants of antiquity. These latter pieces were brass which was considered superior to the iron castings produced by John Browne's Sussex foundry. Browne was a 'Royal gunfounder' the production of artillery being under licence from the King, yet he had been the principal foundry for Parliament since the war began.[242] As during any war, the arms manufacturers were amassing small fortunes from the nation's misery, and John Browne was no exception to this general rule. With foundries at Brede in Sussex, and Horsemonden in Kent, Browne had supplied both land and naval pieces to Parliament. It is likely however, that while the 'fleet' made order separately from the army committee, their provision according to size and loading requirement were identical, making estimates for the navy made on 19 March 1645 of particular interest:

16 demi-culverin, 10 sakers	£428 15 s.
20 saker-drakes, 4 demi-culverin cuts	£416
10 minion cuts	£15 10 s.
Round shot for the several pieces	£1392 17s. 3d.
Bars of iron	£29 17s. 4d.
Hand grenades for demi-culverin, sakers	£125

Demi-culverin and sakers were used as field artillery pieces, and it is worth noting that either could be fitted with grenades to obviously give the bomb greater range. Although these were to be used at Sea, there is no reason to believe grenades were not used in this way on land during sieges, and largely reiterates the old myth that all shot was solid, except for case shot contained in bags and firing chain and nails. Although still primitive compared to the advances made in the late seventeenth, early eighteenth centuries, the artillery of 1645 was not quite the unscientific firepower of the middle ages some people would maintain.

The logistical and technical specifications for artillery were as follows:

Artillery	Calibre (in inches)	Weight of Piece (in lbs)	Length of Piece (in ft)	Weight of Shot (in lbs)
Cannon Royal	8	8000	8	63
Cannon	7	7000	10	47
Demi-cannon	6	6000	12	27
Culverin	5	4000	11	15
Demi-culverin	4.5	3600	10	9
Saker	3.5	2500	9.5	5.25
Minion	3	1500	8	4
Falcon	2.75	700	6	2.25
Falconet	2	210	4	1.25
Robinet	1.25	120	3	0.75

Table 7: Artillery specifications

Of these pieces the Cannon, Demi-cannon and Culverin would be called to sieges from strategic garrisons, whereas Demi-Culverin and particularly Saker were carried in the Train with the army.

The ranges of such pieces could vary with the casting, wind and general climatic conditions, but the theoretical ranges were as follows:

Artillery	Point Blank	10° elevation
Culverin	460	2650
Demi-culverin	400	2400
Saker	360	2170
Falcon	320	1920

Table 8: Range (in yards)

In addition heavy mortars firing large explosive grenades were available for sieges, but required stable, prepared positions from which to ply their deadly trade.[243]

To resolve the technical details of besieging a major city like Oxford, Fairfax called a Council of War. Between them, Fairfax, Cromwell and Brown had Oxford covered on all sides. Cromwell had set up his headquarters at Witcham, Brown at

Wolvercote, and on 26 May Fairfax put over four regiments and thirteen Carriages at the bridge over the Charwell.[244]

The approach of Brown's forces to Wolvercote had caused the King's garrison at Godslow to desert it, putting the house to the torch, although this latter excess was extinguished by a patrol under Colonel Sheffield from Fairfax's advancing force. In addition to saving the house and its outbuildings, a good supply of Powder and Ammunition was preserved for use by Parliaments' forces. Furthermore the garrison commander was taken on the road to Oxford by Sheffield's troopers, and delivered a prisoner.[245] The spring of 1645 can almost be seen as the King's period of military arson, for along with Godstow House the governor torched a Mill. This latter action was prudent military discipline, for Mills were excellent viewing platforms, giving a clear view of surrounding countryside over many miles. Also on 24 May, Fairfax had decided on a small siege at Boarstal House. Adjutant General *Flemming* was there engaged in a single encounter, shot his enemy, yet received a wound himself, conceived then to be mortal, but of which he afterwards recovered.'[246]

Although Fairfax had force-marched his army from Salisbury and was ready to proceed with the siege of Oxford itself, yet the whole matter was thrown away by a further order of 28 May from the Committee of Both Kingdoms, which stated,

> SIR,
> Upon the 25th of this Instant, the King's Army returned to *Tutbury*, and upon *Monday* the 26th, was about *Ashby* intending in probability for the Association; upon which intelligence, we have again written to Lieutenant General *Cromwell* to repair speedily to the Isle of *Ely*, and to take with him four Troops of Horse, and we desire you to give Order accordingly. We have also written to the Committee of *Cambridge* to look to the security of their Garrisons, and to send their Forces to Rendezvous as Lieutenant General *Cromwell* shall appoint. We have written to Colonel *Vermuyden* (unless the King march Northward) to come back towards the Association with all diligence, and to join with such other Forces, as we have appointed, for the defence thereof, and to obey such further directions as he shall receive from this Committee, or from you. We have also written to Colonel *Massie* to come to *Burford*, with what force he can, having regard to the security of his Garrisons, and those parts, and holding Intelligence with you to joyn as there shall be occasion. We desire you without interrupting your proceedings for the present about *Oxford* to have your Forces in such posture as they may be ready to march for opposing the Enemy, and securing of the Association if there shall be need.[247]

During this period, Vermuyden's force, being expected to cover the flanks of Leicester, was the only restriction to the king's advance, and by joining with the

Scots prevented Prince Rupert striking northward. The Scottish army however, far from being prepared to form a field army with Vermuyden, had since withdrawn themselves away from Derbyshire on the pretext of a threat to the borders from the Marquis of Montrose. While in the vicinity of Leicester, Sir Robert Pye left his regiment and began to assess the military defences of that place. Pye's regiment continued with Vermuyden's force, and was commanded by its Major, Matthew Tomlinson, who would see it through the following weeks.[248] Vermuyden's force, drawing off from Leicester, now left the county without field cover, except for the militia regiments in the town itself.

The strategic situation in the rest of the nation was equally as difficult. A week before Fairfax sat down before Oxford, the King had marched from Droitwich in the West Midlands, towards Chester on the North Wales border. In Wales, Sir Charles Gerard had gained a victory over the Parliamentary commander Laugharne, and now threatened the port of Milford Haven. At Scarborough that fine soldier Sir John Meldrum had fallen, and the King's march north now caused the siege of Chester to be raised.

Reaching Market Drayton, the King was joined by the forces under Lord Byron, who brought news from Chester. Meanwhile, old Lord Ferdinando Fairfax, father of Sir Thomas, sent word to Leven to hasten to the town of Manchester in a bid to halt the King's threat against Sir William Brereton's forces, which were now retreating after besieging Chester. Leven on the other hand had received news of Montrose's victory at Auldearn and began to question the wisdom of such a move. In complete panic the Committee of Both Kingdoms rushed orders to their Parliamentary garrison commanders, and ordered their northern officers to virtually wake up to their situation. Leven now agreed to move south, but disclosed his apprehension that Charles intended to join with Montrose.

On 25 May, Edward Massey was ordered to hold Lord Goring in the West Country. Massey had been one of Parliaments better garrison commanders at Gloucester, and with Waller in semi-retirement, would soon be put in overall command of a new Western Association Army (or brigade). Never doing anything by halves, Massey stormed Evesham on 26 May and blocked the intelligence network between Oxford and the King, and standing in direct line of Goring should he attempt to join the King.

The King, receiving intelligence that Leven was now in Westmorland, decided to march towards Scotland via the eastern side of the country, through Yorkshire and perhaps threaten Newcastle-upon-Tyne and the Bishopric of Durham. This held the additional advantage of utilising all the propaganda value of Scottish control being in northern counties, and surely good Englishmen would fly to his banner if he went to restore English soil to Englishmen. In Yorkshire, a county where being born in the wrong town can mean one's a foreigner, the Earl of Leven's forces were not popular due to the over eagerness of his troops to 'liberate' anything worth 'liberating' and Charles had since 1642 found recruits there.

If the King was to march on Scotland, he would nevertheless need every soldier available, and with this in mind he sent direct orders for Gerard and Goring to join him on the march. Almost at once, news that Oxford was unable to withstand the siege due to lack of provisions reached the King. The orders to Goring immediately changed, he was now to march by way of Newbury and by threatening London draw the New Model away from Oxford. This did not however take Massey into account. If this did not work - as it would not - then the King would be forced to return south, join with Goring and relieve Oxford himself.

All in all, the King had given himself too many options. While in this openminded state, Prince Rupert put another plan forward, a quick strike at Leicester, from where the Royalists could march either way.

The drawing away of Vermuyden's forces into the so-called Association areas left Leicester unprotected, and on the morning of 28 May the Parliamentarians awoke to the sight of smoke rising from the windmills in the countryside. Sir Marmaduke Langdale the commander of the King's northern horse had been sent forward by Rupert to secure the outer roads to Leicester, and had fired the mills, almost by force of habit. It was no coincidence that Sir Robert Pye was in the town, the Committee of Leicester having allowed the defences to fall into terrible disrepair, few of the outer walls having linings to reduce artillery damage. Pye was a capable if not inspired commander of horse, and had had three years experience in the field, however his greatest asset was probably in military engineering and the defences required to withstand a siege. What Pye found at Leicester was this woeful neglect of the old walls, while no significant earthworks had been built to extend the defences into the fields around the town, and it was now a matter of too little, too late. Admittedly, after nearly three years of civil war passing them by, the people of Leicester were very much still in a cold war situation, and after all, why spend hard earned money on a war that had never really arrived? They would pay a dear price for that folly now.

To test the strength of the royalists and push back the 'pickets' or forlorn hope, Colonel Pye sent out a body of horse commanded by militia officers, Major Innes and Captain Babbington, to drive hard and swift at the Royalist lines. That same afternoon Lieutenant Davies charged another advance party of Royalists and chased them as far as Belgrave Bridge. The King's artillery pieces now began to arrive, and the opening cannonade ripped into the walls of Leicester. Pye called for gabions, tall baskets filled with earth as protection for the towns own gunners, there were none to be found.

The following day saw little action. Instead the Royalists directed their efforts to constructing batteries for their heavier guns. The Parliamentarians played on these positions with musketry and their own artillery, but the lighter guns in Leicester did little damage. On the morning of Friday 30 May, the Commander-in-Chief of the Royalist forces, Prince Rupert, made his first offer of terms. In perfect keeping with Rupert's character he 'shott two great pieces' into the town after which he sent forward a Trumpeter with an offer to allow the 'Horse to march out', if the garrison surrendered immediately. The Trumpeter was held in Leicester while the Corporation decided what to do. At the same time Robert Pye held a Council of War. Pye and Major Innes, both experienced soldiers, advised the Corporation to accept the prince's terms, knowing full well that Leicester could not withstand a full scale assault by the King's main army. Nevertheless, the Committee headed by Colonel Grey and the militia Captains, Babbington and Hacker, sought to defend the town, one of them, probably the blunt and stubborn Francis Hacker, declaring 'We are part of those who have undertaken the Parliament's cause, a cause so high as I desire to die in no other.' Rupert in the meantime began to 'rayse a battery for six great pieces upon a hill, where sometymes of old had byn such another'. The Committee realising that they could not prevent an assault, now tried to buy time, asking why Rupert was preparing siegeworks during a period of truce. At this impertinence, Rupert flared, and as the royalist officer Richard Symonds reports 'His Highness told the trumpet [sent by the Committee] if he came agen with such another errand he'd lay him by the heels'. Shortly afterward, the same Trumpeter took the same message to Prince Rupert, who this time imprisoned him with the Provost Martial. Prince Rupert was furious. He had been humiliated at Leicester in 1642, and this same town was now refusing what he considered very favourable terms. About three o'clock that afternoon, the Royalist guns began the cannonade upon an 'unlyned' stonewall on the south side of the town. If this action did nothing more it ended the meeting of the Leicester Common Council who were trying to decide whether to fight or not. Needless to say there is nothing like a cannonball to hasten deliberation.

The question however remained, would the soldiers in the town fight? Many were of poor quality, some were Scots who had trickled into the town following their refusal to serve in the New Model. To bolster their resolve, members of the Council went from soldier to soldier handing out money, a paid soldier no doubt being less likely to run.

Within three hours the wall near Newark Gate was breached, but resistance was strong. Richard Symonds relates that 'after the breach was made in the wall by cannon, by six of the clock, they in the towne had gotten up a handsome retrenchment with three flankers, (a great Spanish piece) within four or five yards of the wall.'[249]

The people of Leicester showed great courage in building an inner position from which to fight. The cannon fire continued for the next six hours, the width of the breach growing all the time. Under the cover of darkness, just before midnight, Prince Rupert ordered the storm to begin. Assault by storm was the most violent form of siege, with no quarter to be expected, simply to take the town as quickly as possible being the order.

The Royalist foot pressed on, attacking the breach, and at Belgrave gate Sir Henry Bard fell on with scaling ladders. On the north side of the town Sir Bernard Astley sent his men against the drawbridge near St. Margaret's Church. On the south side, George Lisle led his men against the breach, the defenders waiting until the royalists entered and then galling them with case shot from the flanking guns. Despite a very gallant defence the Parliamentarians were forced back, but suddenly Sir Robert Pye sent Francis Hacker with a squadron of horse to charge the foot, ably backed by dragoon fire commanded by Major Innes. Pye understood that in the narrow streets a few well-armed troopers could undertake the work of a regiment in the field. The Royalists at the breach fell back.

The King himself hearing of the difficulty, gave order that his own Lifeguard should enter the breach. The Lifeguard was the cream of the foot, and pressing on at the push of pike, began to make advances against the hastily formed barricade. Attacking one of the guns with desperate valour, the Royalist Colonel St. George was slain when he was blown to pieces at the cannon's mouth. On Parliament's side, a German dragoon, possibly one of the many religious refugees who had fled from Europe during the Thirty Years War, claimed his place in history, when he volunteered to man the guns after the principal gunner was killed. This brave German continued to fire, until finally 'overrun', he was transfixed by a pike to his gun.

At Belgrave gate in the meantime, Sir Henry Bard led his men on, crying for them to follow him. Bard himself climbed one handed up the rough scaling ladder which was the only means of entry along this wall, but having lost an arm at Cheriton in 1644 he was unable to defend himself and received a glancing blow from a musket upon reaching the top. Bard survived, but sixteen of his men fell in this assault with a further sixty wounded. Not deterred, Bard called for grenades, and again fell on, this time tossing the primitive hand shells with great effect. The defenders along Belgrave gate fell back cut and burned, the royalists were in to secure a foothold, whilst reinforcements poured over the walls.

It was now about one o'clock, and fires were breaking out following the grenade attack, and illuminated the night sky. The Belgrave gate flew open, and immediately the Earl of Northampton's experienced cavalry galloped across the drawbridge, cutting at the poor bewildered Parliamentarians, while Colonel Page from the royalist garrison at Newark in Nottinghamshire, dismounted his troopers and with sword and pistol cut his way through the streets.

In the meantime, Sir Bernard Astley's men gained North Mills, a position of strategic importance, and raised the ladders against the works near St. Margaret's. Colonel John Russell at the head of Prince Rupert's bluecoats, and his red coated firelocks, gained the walls, but reaching the top of his ladder Major Bunnington a gentleman pensioner and Major of the firelocks, was 'shott in the eye' and fell dead.

Having disregarded Pye's advice, the military governor of Leicester, Colonel Theo Grey, now assembled a last body of horse and charged. Cut twice in the face and a pike being driven in his back, Grey was overpowered and received quarter.

At the main battery, possibly at Horsefair Leas, Prince Rupert's black colours flew in victory, but though every wall was taken, the fight continued. Hurrying to the market place the people of Leicester threw up a new barricade of carts, barrels and even furniture from their houses. This defence held out for nearly an hour against terrible odds.

Meanwhile, in the suburbs, the royalists fell to plunder, with sporadic musket fire from the houses forcing them back to fighting from street to street, house to house. The townspeople hurled stones and tiles from the roofs at the royalists below. In St. Martin's churchyard a part of the old garrison still held out behind another defence line. The Parliamentary horse under Captain Babbington made a fierce charge driving the royalists towards the Market Place, where despite cavalry support the soldiers of the garrison finally surrendered. By taking Leicester by storm, Rupert had quickly ended what could have been a long drawn out siege, and for the Parliamentarians is had been at a terrible cost. The Scots were killed almost to a man, likewise women who had aided their men folk at the barricades. It is said that every street had its dead, a fact not denied by any royalist chronicler.[250]

The fall of Leicester was almost certainly due to the removal of Vermuyden's forces from the area. This allowed the King's horse a free hand to secure the approaches towards the town, and the foot to march up. Leicester had been the cost of the Committee of Both Kingdom's decision to secure the Eastern Association. It was a disaster of the logistics, which had born the New Model, yet was no fault of the General or the soldiers in his army. The strategy was in chaos and Cromwell, who was the only other force within marching distance of Leicester, was left idle. The diarist D'ewes concluded, 'that the first error was the calling back of Colonel Cromwell and now it be fit to know where Colonel Vermuyden is and whether the Scots' forces be advanced.'[251]

A letter from Sir John Norwich at Rockingham Castle to Sir Samuel Luke at Newport Pagnell fills in more detail:

> Since my last I am sorry it is my misfortune to give you intelligence of the great loss of Leicester, which the enemy, Friday towards break of day, desperately stormed in many places at once though bravely repulsed by the horse and foot after their entrance into the town, wherein were taken prisoners Sir [Robert] Pye, Major Ennis, Capt. [Francis] Hacker, Capt. Babbington, the Committee and what other commanders of quality I know not. The King, Rupert, Maurice and Langdale all there with 7,000 horse and 4,000 foot.

> Colonel Vermuyden and Colonel Rossiter appointed to meet Nottingham and Derby horse on Saturday at six in the morning at the rendezvous at Melton Mowbray, thinking thereby to draw off the enemy, but Nottingham and Derby came not. Colonel Vermuyden and Colonel Rossiter, being at Melton Mowbray, having certain intelligence of Leicester being taken, made their retreat to Grantham and are quartered thereabouts. Which way they intend I know not. Kirby and Butleigh have both quit their garrisons. Butleigh have left ordnance and ammunition and other provisions behind them. I have sent out a party of horse to recover them if it be not too late.[252]

Although Vermuyden had hitherto been in command of the New Model forces with him, once in Lincolnshire he found himself commanded by Oliver Cromwell who had received orders to command all the forces in that area.[253] It is interesting that Cromwell, not a member of the New Model, was given precedence in command. This is not without justifiable reason, Cromwell was still a Lieutenant General despite the terms of the Self-Denying Ordinance, but under normal circumstances an officer of the Associations would not supersede one of the regular army. This would therefore suggest, that the terms of engagement being administered by the Committee of Both Kingdoms, were indeed abnormal, in so much that to the Committee the security of strategic areas was of paramount importance, not strict adherence to the rules so painstakingly argued over between Lords and Commons. By 4 June, the force with Vermuyden was estimated to be '4 or 5,000 strong' although such a figure appears rather high.[254]

Within two weeks Parliament had seen part of its new army trapped in Taunton and its major garrison at Leicester fall to the King. If the spring campaign was not to become a complete disaster, these calamities had to be reversed. Therefore, Sir Thomas Fairfax abandoned the siege of Oxford on 4 June and ordered his army to shadow the King wherever he lay. But the fact remained that on this day, Fairfax simply had no army to shadow the King with, and certainly not an army able to fight a grand field battle. The destiny of the Parliamentary cause lay in what could be done in the next ten days. This period alone would be the watershed that shaped the future.

NOTES

[236] John Rushworth, *Historical collections of private passages of state* (7 vols., 1659-1701) part III, vol. 1, pp.32-33.
[237] Ibid. p.33.
[238] Ibid.

[239] Ibid. pp.33-4.

[240] Heavy siege artillery and heavy mortars rarely travelled with the field army, they were invariably transported as required from garrisons within marching distance.

[241] *C. S. P. Dom. 1644-45*, p.516.

[242] Sir Charles Thomas-Sanford, *Sussex in the Great Civil War and Interregnum 1642-1660* (1910) p.176. In June that year, Browne was implicated in assisting the Kent royalists, ibid. pp. 176-178.

[243] These tables come from P. Young, *Edgehill 1642, the campaign and the battle* (Kineton, Warwickshire, 1967) pp.31-32. It is true that other scientific tables exist giving highly detailed mathematical specifications, but as Peter Young once told me "none of them were tested in a damp English field with a gale blowing".

[244] One must assume that the four regiments were of Foot, and the carriages were light field artillery.

[245] Joshua Sprigg, *Anglia Rediviva, England's Recovery* (1647) p.22.

[246] Ibid.

[247] Rushworth, *Historical collections*, part IV, vol. 1, p.34.

[248] Whether Pye took his own troop with him into Leicester is uncertain, but Joshua Sprigg on his battle plan of Naseby drawn by Streeter, shows only five troop cornets for Pye's regiment, signifying the Colonel was absent. For further details on Pye's regiment see Chapter XI below.

[249] R. Symonds *The Diary of the Marches kept by the Royal Army during the great Civil War*, ed. C. E. Long (Publications of the Camden Society, 74, 1859).

[250] This section on Leicester is based on Rushworth, *Historical collections* part IV, vol. 1, p. 35; R. Symonds *The Diary of the Marches kept by the Royal Army during the great Civil War* ed. C. E. Long (Publications of the Camden Society, vol. 74; 1859); Clarendon, Edward Earl of, *The history of the rebellion and civil wars in England*, ed. W. D. Macray (6 vols., Oxford, 1888) book IX, vol. 4; but primarily on the author's own article 'The Eye of the Storm - the siege of Leicester 1645' B. Denton (Leicester Graphic, February 1985).

[251] 'Diary of Sir Simonds D'ewes', BL Harleian MS 166, f.215.

[252] *The letter books, 1644-45, of Sir Samuel Luke, Parliamentary General of Newport Pagnell*, ed. H. G. Tibbut (Publications of the Bedfordshire Historical Record Society 42; HMC joint publications, 4, 1963) p.552.

[253] Ibid. p.551.

[254] Ibid. p.556.

CHAPTER XI
THE ROAD TO NASEBY FIGHT

The fall of Leicester had highlighted the deficiencies in the strategy of the Committee of Both Kingdoms, but at the same time news reached London of Weldon's dire condition in Taunton. The forces of Goring, Sir Ralph Hopton and Sir Richard Grenville had joined together, and although Weldon could hold the royalists in a defensive position, his brigade could no longer march back to Fairfax.

In the meantime, Cromwell had been sent with four Troops of Horse to the Isle of Ely. This was Cromwell's home country, his house being in the old cathedral town of Ely, and his Parliamentary constituency being nearby Huntingdon.

On 31 May, Fairfax ordered Colonel Thomas Rainsborow with his regiment and a body of Horse to besiege Gaunt House, which upon being summoned, it surrendered the next day.[255] During the time Fairfax was before Oxford, the garrisons, which formed the King's outer circle of defences, were slowly being prevented from receiving fresh supplies. To ease the situation, the Governor of Oxford, on Monday 2 June, sallied forth as Rushworth recalled:

> about one of the Clock in the Morning marching himself with near 1000 Horse and Foot towards *Hedington*-Hill (a mile from *Oxford*) where there was a Guard kept by the Besiegers; and sending Col. *Walters*, Sir *Thomas Gardiner*, and Capt. *Grace* towards *Bullington-Green*, to fetch a compass, and fall in upon them behind; the whole Guard at that place was surprized, some kill'd, and fourscore and twelve carried away Prisoners into *Oxford*, whereof one was Capt. *Gibbons*; but the next day they were Released in Exchange for as many of those who had been taken in the Skirmish between Adjutant General *Flemming* and Capt. *Gardiner* before mentioned.[256]

There were two Captains by the name of Gibbon or Gibbons in the New Model Army, a Captain of Foot under Philip Skippon, and Robert Gibbons of Ireton's regiment. Skippon's regiment had however, been laying siege to Boarstall House for some days, suggesting that Gibbons was the Kent officer now under Ireton, who must have rendezvoused with Fairfax sometime after 15 May when the rest of the regiment rode north under Vermuyden.

The following day on Wednesday 4 June, Sir Thomas Fairfax lifted the Siege of Oxford after receiving direct orders to defend the Association. Fairfax immediately gave orders that the bridge the New Model had thrown over the river at Islip was to be pulled down, and for the forces on that side of Oxford to rejoin him on the march.[257] On this march the General turned out of his way to see Skippon besieging Borstal House. This was a substantial garrison for the King, which had stubbornly refused terms from Skippon. Needing to release the forces lying before the garrison, Fairfax sent a further summons to the military Governor, Sir William Campion:

> SIR,
> I send you this Summons before I proceed to further Extremities, to deliver up to me the House of *Borstal* you now hold, with all the Ordnance, Arms and Ammunition therein, for the use and service of the Kingdom, which if you shall agree unto, you may expect Civilities and fair respects, otherwise you may draw upon your self those Inconveniences which I desire may be prevented. I expect your Answer by this Trumpet within one hour. I rest,
>
> Your Servant,
> *Tho. Fairfax.*

There is no reason to believe that Sir William Campion had received intelligence of the fall of Leicester, and every reason for him to believe Fairfax was still besieging Oxford. Yet the Royalist gave the following staunch reply:

> SIR,
> You have sent unto me a Summons to Surrender this House, for the Service of the Kingdom; I thought that Bait had long e'er this very stale, (considering the King's often Declarations and Protestations to the contrary) now sufficient only to cozen Women and poor ignorant people: For your Civilities as far as they are consonant to my Honour I embrace; in this place I absolutely apprehend them destructive not only to my Honour, but also to my conscience: I am therefore ready to undergo all Inconveniences rather than to submit to any, much less to those so dishonourable and unworthy Propositions.
> This is the Resolution of,
> Sir, your Servant,
> *W. Campion*[258]

That night Skippon attempted the House by Storm, but the garrison withheld the assault and rather than divide the army still further, Fairfax withdrew these forces too.

On 4 May the Committee of Northampton wrote twice to Fairfax with news of the King's march. That day Sir Robert Pye who was taken by the King at Leicester, had after just four days been exchanged for Sir Henry Tillier who had fallen prisoner at Marston Moor the year before. Pye still on 'parole' and therefore debarred from serving until the period elapsed, made straight for Northampton to warn that Committee that the Royal army headed that way, obviously with the intent to render that town under storm as it had reduced its near county neighbour. Unlike at Leicester, the Committee and townspeople of Northampton had kept their fortifications in good repair, with strongly maintained walls, well-guarded approaches, and the rivers of the county adding substantial obstacles to an advancing army. At 10 o'clock at night that same day, the Committee wrote their second, more urgent report to Fairfax:

> SIR,
> We have at this Instant received certain Intelligence that the King's Army is advanced this way, and a great party both of Horse and Foot come as far as *Harborough*, which is within twelve Miles of us; their design is thought to be for this place; we doubt not but you will take us into your thoughts for a speedy Relief, and in the mean time we shall provide for them to the best of our strength. We Remain,
> Your Humble Servants,
> Edward Farmer
> Rowland St. John
> E. Attarby
> *Richard Samwell*[259]
> *Philip Folman*5

Early on 6 June a portion of the King's army faced Northampton, but with news of Fairfax on the march it was not prudent for Prince Rupert to attack the town.[260]

Meanwhile, Cromwell had installed himself at Huntingdon where he wrote to Fairfax on the 4th, relating the poor conditions he had found in the Isle of Ely. He also reported that Vermuyden's forces had reached Market Deeping, with orders to join with Sir John Gell's Nottinghamshire horse if so ordered. In the same letter Cromwell continued with a personal recommendation to Fairfax:

> We heard you were marching towards us, which was matter of rejoycing to us. I am bold to present this as my humble suit, That you would be pleased to make Captain *Rawlins* this Bearer, a Captain of Horse; He has been so before, was nominated to the Model, is a most Honest Man: Colonel *Sidney* leaving his Regiment, if it please you to bestow his Troop on him, I am confident he will serve you faithfully.[261]

Fairfax did not in fact give Sidney's old troop to Rawlins for this had passed to Nathaniel Rich, but with Sir Robert Pye being effectively removed from the army by his parole, Rawlins was added to this regiment while it was commanded by Tomlinson.[262]

If the King's forces were to be prevented from overrunning the Association, or generally gaining a military advantage from the lack of coordination between the divided Parliamentary army, then the Royalists simply had to be checked in the field. In the first week of June 1645, the Parliamentary forces were logistically in a worse position than they had been at Newbury the previous autumn. If in desperation the hawks in the Commons and Lords had forced through all the constitutional changes required to mould the New Model, it was hardly conducive to their military or political good to find their warlike offspring incapacitated by poor strategy.

To bring the King to battle, Fairfax needed to reassemble his army. To this end Vermuyden was ordered to rejoin Fairfax, and wrote to the General on 6 June:

> *May it please your Excellency,*
> Last Night upon Intelligence of the King's March towards *Northampton* I came to *Oundle*, and am now marching to meet your Excellency, who I hear is about *Brackley*. The Lord send us a speedy Conjunction, a good Engagement with blessed success. Which is the earnest Prayer of,
> Your Excellen*cy's most Obedient*
> *Faithful Servant,*
> *Vermuyden*[263]

At the same time Bartholomew Vermuyden wrote separately to Samuel Luke at Newport Pagnell that 'upon intelligence of the King's march southward I have hastened to join Sir [Thomas] Fairfax who I hear is about Fenny Stratford. I am with my brigade at Oundle, Weston and Laundon. Let me know by this bearer where I shall send to him for orders.'[264] The Committee of Both Kingdoms had ordered Vermuyden's mounted brigade to rejoin Fairfax's main force which was shadowing the Royalist main army between Oxford and Northampton. In the meantime, Fairfax had reached Brickhill, where he set up his headquarters on the 6th. During this stay Sprigg tells of a sudden and potentially dangerous occurrence:

> This night a great fire happened at the Generals quarters at *Brickhil*, which was so sudden and violent for the time, that a man and a boy, and three or four horses were burnt in the Barn where the fire began, before the Guard could get to preserve them. It happened most remarkably, in the house of one who expressed no good affection to this Army, and denied to furnish those conveniences for quarter, (affirming that he had them not) which afterwards by occasion of the fire, he was enforced to bring out.[265]

The Road To Naseby Fight

On 7 June, Fairfax marched the army from Brickhill to Sherrington, a mile east of Newport Pagnell.[266] Joshua Sprigg relates that,

> The next day June 7, the Army marched to Sherrington, a mile East of Newport-Pagnel, to the end the Forces with Colonel Vermuyden (who upon the Scots retreat to Westmorland were recalled, and upon their march back) might more conveniently joyn, but especially to be on that side the River, the better to secure the Association, in case the King, who the day before had faced Northampton, and seemed to intend that way.[267]

There can be little doubt that the news of the fall of Leicester had reached Vermuyden within a day or two of the disaster, and with battle looking imminent the regiment rejoined the New Model on 8 June. Fairfax had the previous day written to Cromwell with detailed instruction how to join with the New Model should the King march into the Association, and on the 8th the army patrols rode into Northamptonshire as far as Towcester.

The arrival of Vermuyden back with 2,500 horse strengthened Fairfax's forces considerably, but little could have prepared him for the Dutchman's action soon after the rendezvous. Whether the influence of presbyterianism and the political aspects of the Covenant had any effect upon Vermuyden is unknown, but during 1644 the arguments between the presbyterian faction led by the Scottish soldier Crawford, and the army sectaries if not epitomized, certainly supported by Cromwell, had caused a bitter friction in the Eastern Association forces. On that same 8 June, Vermuyden called upon Fairfax as Sprigg again relates:

> This day, Colonel Vermuyden, who the day before was with his party of Horse returned, and come near to the quarters of the army, himself came to the General, desiring [in regard of some speciall occasions which he said he had to draw him beyond seas] that he might have leave to lay down his Commission, which was yielded unto, and accordingly he received his discharge.[268]

Sprigg gives a small insight that the real reason for Vermuyden's resignation was different to the personal problems claimed by the Colonel, by saying 'in regard of some special occasions which he said he had to draw him beyond seas'. John Rushworth writing of the same resignation, and probably referring to Sprigg's original hand written text, supplants the words 'which he said he had' with the stronger 'he alleged', which must add considerably to the speculation to the reason for Vermuyden's departure. At a time when the New Model which was designed to reverse the military inertia of autumn 1644, and wipe away the discourse between factions, the loss of an experienced officer, like Vermuyden, was all Sir Thomas

Fairfax needed. The loss of a potential Lieutenant-General of Horse did however give Fairfax the perfect opportunity to request that Oliver Cromwell should be given further special dispensation from Self Denial, to join the New Model, and he wrote that same evening to the Lords with that very request. According to Rushworth, the letter was written by Colonel Hammond, a kinsman of Cromwell.[269]

It is interesting that Fairfax sent this letter of Sunday 8 June 1645, the Sabbath, when no Parliament sat in session to accept correspondence, yet his letter must have arrived to be heard early on the 9th, and the order passed through the procedural records of the Committee of Both Kingdoms on the 10th. Although Cromwell was ordered to join Fairfax with a body of horse and dragoons from the Associations, it was not until late on the 11th that he received further orders from Fairfax:

> SIR,
> You will find by the Inclosed Vote of the House of Commons a liberty given me to appoint you Lieut. Gen. of the Horse of this Army during such time as that House shall be pleased to dispense with your attendance, you cannot expect but that I make use of so good an advantage as I apprehend this to be to the publick good. And therefore I desire you to make speedy repair to this Army, and give order that the Troops of Horse you had from hence, and what other Horse or Dragoons can be spaired from the attendance of your Foot in their coming up, march hither with convenient speed; and as for any other Forces you have there, I shall not need to desire you to dispose of them as you shall find most for the publick advantage; which we here apprehend to be that they march toward us by the way of *Bedford*. We are now Quartered at *Wotton* [Wootton] two miles from *Northampton*, the enemy still at *Daventry*. Your Intelligence is that they intend to move on Friday, but which way we cannot yet tell. They are as we hear more Horse than Foot, and make their Horse the Confidence. Ours shall be in God, I pray make all possible haste toward.
> Your Affectionate Friend
to Serve you,
Tho. Fairfax[270]

For what was supposedly a last minute life or death request to add a Lieutenant General to the ranks ready for the test of battle, Oliver Cromwell's movements, added to Sir Thomas Fairfax's need, and Vermuyden's willingness to vacate the Army were all most convenient. A more cynical chronicler might well speculate on the point to be made in the future by Cromwell's less friendly contemporaries, that the whole situation was pre-planned, and he was always in a position where he would retain, or reassume the command in the New Model. By Cromwell joining the new army, the power balance could be slowly weighted in favour of the Independents. It is true however, that Oliver Cromwell was the military find of the civil war, in comparison with his own generation he was more than anyone's equal

The Road To Naseby Fight

upon the battlefield, despite having no military experience prior to 1642. Even compared with Prince Rupert he had a wisdom in his decisions not found in the more outwardly militaristic prince.

It is interesting to note that although the ordinance for Self Denial barred members of Parliament from service in the new army, there was no provision to enforce the same. Cromwell simply required the permission of the Commons to authorize a commission from Fairfax, no further ordinance was needed to release him from Self Denial. A similar problem would have arisen even if Cromwell had tried to resign his seat in order to retain his military command. Because he had been returned to represent a franchised town in Parliament, he could not stand down freely from that responsibility and trigger a by-election.[271]

The quarters being sparse in Wootton, Fairfax and his principal officers rode into Northampton, where the Mayor, Joseph Sergeant, and the town's magistrates arranged suitable accommodation, and that night visited the General at his headquarters.[272]

During the march from Oxford, Major General Skippon was ordered to drawn up the 'form of a battel', this being the order for deploying the foot in the field. Skippon had eight regiments at his disposal, those of Fairfax, Philip Skippon, Hardress Waller, Robert Hammond, Edward Harley (commanded by Pride), Edward Montague, John Pickering, and Thomas Rainborowe. Five of the eight regiments had their roots in the army of the Earl of Manchester's Eastern Association, two in Essex's army, and the remaining single regiment from Waller's. Added to these seasoned veterans were the conscripts, or pressed men, from the counties. Skippon therefore had had less than two months in the field to train and mould these eight regiments into a fighting force.

With the return of Vermuyden's brigade, and the small detachment of Association horse under John Fiennes in addition, the New Model cavalry was only two regiments short of establishment strength by 8 June, the two missing units being Rossiter and Graves. Upon his coming therefore, Cromwell would have the regiments of Fairfax, John Butler, Thomas Sheffield, Charles Fleetwood and Bartholomew Vermuyden (commanded by Huntingdon), Nathaniel Rich, Sir Robert Pye, (commanded by Tomlinson), Edward Whalley and Henry Ireton at his command. In addition, the small Life-guard under Charles D'Oyley, the troops under John Fiennes, and the horse to be spared out of Ely would also be at his disposal.

In a last attempt to swell his forces still further, Fairfax sent messengers to Sir John Gell, Colonel Rossiter, and the military Governors of Coventry, Warwick, Northampton and Nottingham, to watch over their own defences, but to send what troops they could to reinforce his own army. On Wednesday 11 June, Fairfax sent a full report of his condition, and the whereabouts of the King to the Committee of Both Kingdoms:

> The Army is this day marched from *Stonistratford* to *Wotton*, two miles from *Northampton*, and about nine from *Daventry*, where the Enemy yet remains. The ill Weather hath been some disadvantage to us, but I hope a fair day may recover what a foul day loses. It is intended that the Army move something nearer the Enemy to Morrow, by which its like we may discover their resolution. If they appear forward to engage, I shall take the best advice I can upon the place, but as yet I see no great reason to decline. But if the Enemy retreat Westward, or stand upon their advantage till further Supplies, which I hear they expect from the *West* to come to them, I offer it to your Lordships, whether it be not fit that this Army receive what addition conveniently they may. And seeing we lye at as good an advantage, for marching Northward as the Enemy, being nearer to *Newark* their only Pass, I think fit to propound that the Horse of *Lincolnshire*, *Derby* and *Nottingham* be drawn this way with all convenient speed, for if the King go toward the *West*, or the Western Horse come to him, we shall be much inferior in Horse, being already over-numbered. I have already sent to those Horse, but thought fit not only to give your Lordships notice, but desire your assistance in the accomplishing of this motion which appears but reasonable.[273]

The weather of May and June 1645 had been particularly wet, and would have reduced many of the less well-kept roads to a muddy track. The more obvious drovers roads had, before the war, received regular maintenance, but many were now in disrepair, making the movement of artillery and ancillary carts a slow and often arduous task. Fairfax was quite correct to be anxious about Goring's Western horse joining the King, for they were sufficient in number to overpower the New Model cavalry when added to the King's, Rupert's and the Newark and Northern horse now around Daventry. There was no way that Fairfax could have known that despite repeated orders, Goring was still before Taunton, therefore he was exhibiting prudence in both drawing the royalists into premature tactical decisions, whilst assembling as many Association soldiers as possible. It is however noticeable that in this letter, Fairfax was less subservient to the Committee, and indeed informed them of his orders to the counties. This was a distinctive alteration in the relationship between General and executive, for Fairfax had not previously set his own strategy. It is interesting to ponder, that once Cromwell had been named Lieutenant General, the inner confidence of Fairfax appears to have grown.

The following day, on 12 June, the Army marched to Gilsborough, four miles west of Northampton, and five miles from the royalists on the slope of Burrough Hill. A Parliamentary patrol sent in search of the enemy was engaged in a skirmish and took a small number of prisoners. From these it was understood that the royalist army was encamped around the ancient defences of the hill, with the horses being allowed to take on condition by being allowed to graze to green summer grassland.[274] The King at this same time, amused himself with hunting the deer in Fawsley Park.[275]

That night one of the stranger occurrences happened in Fairfax's career when, being unable to sleep, he mounted his horse. Rushworth recorded how Fairfax,

> rode about the Horse and Foot Guards till Four in the Morning, where an odd Adventure happen'd: Having his thoughts otherwise busied, he himself forgot the Word, and was stopt at the first Guard, whereupon declaring who he was, and requiring the Soldier that stood Centinel to give it him, the Fellow refused, saying, He was to demand the Word from all that past him, but to give it to none; and if he advanced without it, would shoot him. And so made the General stay in the wet till he sent for the Captain of the Guard to receive his Commission to give the Word; and in the end the Soldier was rewarded for his Duty and Carefulness.[276]

Such pieces almost take on the trappings of folk tales with their telling, yet the story serves to illustrate the uneasy mind of the General, and furthermore that all history does not hinge upon solely the deeds of great men. About three o'clock that same morning, Fairfax continuing his eventful ride to near the village of Flore, he could see the royalists riding away from Burrough Hill, and clouds of smoke rising from the infantry camp. With Fairfax so near, and twelve hundred horse having returned that same morning from Oxford, the King ordered his army to decamp from Burrough Hill, and accordingly sent his carriages toward Market Harborough, the soldiers burning their rough huts in the fields where they had camped.[277] Upon the General's return to his headquarters in Gilsborough, he was attended by Scoutmaster-General Watson, who Sprigg describes and diligent in his work, to confirm what Fairfax had seen with his own eyes. Watson,

> brought him certain notice, that the Enemy was drawing off from *Burrough hill,* had stood in arms all night, and were all amazed that our Army was so near, it being spread abroad in their army we were gone for security into the *Association*; And four or five more of the Spies came one after another, confirming the same intelligence...and this unexpected march of the Army close up to them; caused them speedily to resolve upon their forementioned march towards *Pomfract*; either judging the Army would not follow them, or if they did, they should be able to fight us at more advantage, after they had drawn us further Northward.

At six o'clock on 13 June, Fairfax called a Council of War. In the middle of their debates, cries went up "Ironsides...Ironsides has come", it of course heralded the arrival of Cromwell.[278] He had ridden with the utmost haste out of the Association counties, with six hundred horse and dragoons. With Cromwell's coming the battle plans moved on apace, drums were ordered to beat to assemble the Foot, and trumpets called out for troopers to horse, the whole army to rendezvous at a small market village called Naseby. In the meantime Thomas Harrison with a squadron from Fleetwood's regiment was sent scouting towards Daventry, with a further patrol under Ireton heading by way of Naseby. Sprigg noted how,

> another strong party of Horse was sent under the command of Colonel *Ireton*, to fall upon the flank of the Enemy, if he saw cause; and the main body of our Army marched to flank the Enemy in the way to *Harborough*, and came that night to *Gilling*, the Country much rejoicing at our coming, having been miserably plundered by the Enemy; and some having had their children taken from them and sold before their faces to the Irish of that Army, whom the parents were enforced to redeem with the price of money. That evening we understood that the Van of the Enemies army was at *Harborough*, the Rear within two miles of *Naseby*: and no sooner was the *General* got to his quarters, but tidings was brought him of the good service done by Colonel *Ireton*, in falling into the Enemies quarters, which they had newly taken up in *Naseby* Town; where he took many prisoners, some of the Prince's Life-guard, and *Langdale's* Brigade, and gave a sound alarm throughout the Enemies army.[279]

The auctioning of the children for 'prize money' is an interesting story, although it was probably pure propaganda aimed at Irish involvement with the King's cause.

The main body of the New Model was by dark on 13 June at Gilling.[280] The King's response to the Fairfax's closing with his forces was immediate, calling a Council of War at Harborough where Prince Rupert was quartered, his Majesty put the question whether to engage the New Model or retreat back to Leicester. The Prince, being a man swift to fight, called for action on the grounds that to fall back could endanger the whole army to attack while still divided. Therefore the resolution was to offer battle, 'taking themselves to be more strong in Horse, than Fairfax; to be much better furnish'd with old Experienced Commanders, and having no reason not to relye upon their infantry; for indeed they were generally valliant stout Men.'[281]

With three hundred years of hindsight, it is possible to hypothesize on whether King Charles should have retreated away from Fairfax's stronger force, but it must be considered that he considered the New Model Army to be raw, still below its establishment strength, and untested against a field army of the experience of his own. If the King was to offer battle to Fairfax at any time, it must be now, before the moulding of this New Model was complete. At least the King now had a well-fortified garrison at his back since the storm of Leicester, and this could offer a stop to Fairfax should the tide of events turn suddenly in favour of the New Model. To advance towards Fairfax, and offer battle the following day, was therefore the correct decision. Quoting from Sir Edward Walker, the Earl of Clarendon relates the Royalist position as the early morning light filtered across the fields around Naseby:

> And so, in the morning early, being Saturday the 14th of June, all the army was drawn up, upon a rising ground of very great advantage, about a mile south from Harborough, (which was left at their back) and there put in order to give or receive the charge...the army thus disposed in good order, made a stand on that ground to expect the enemy.[282]

By dawn on Saturday 14 June 1645, the Royalist army was deployed in battle order along the line of Dust Hill to the north of the valley known as Broadmoor. For the defence of his own army, this position was ideal for the King. By simply occupying this line, any advance by Fairfax would expose the New Model foot to a long open attack march in full battail, and limit the charges of the enemy horse against the rise of the hillside.

Parliament's army had likewise been on the march most of the night. Fairfax could not allow the King to escape him for political reasons as well as military, and with the additional horse under Cromwell at his disposal, plus the force formerly commanded by Vermuyden now joined with him, it was in many ways vital to use this force while still available. As we have seen, the danger of waiting was that Goring would join with the King. Therefore, for totally opposing reasons, both commanders had good military considerations to enforce the battle. Joshua Sprigge takes up the story from Parliament's side:

> The General [Fairfax] with the Army advanced by three of the clock in the morning, from Gilling towards Naseby, with an intention to follow close upon the Enemy, and [if possible] retard their march with our Horse, till our foot could draw up to them, in case they should have marched on to Leicester (the intelligence being, that they had drawn some Carriages in the night through Harborough) that way. By five in the morning, the Army was at Rendezvouz near Naseby, where his Excellency received intelligence by our Spies, that the Enemy was at Harborough with this further, that it was still doubtful, whether he meant to march away, or to stand us. But immediately the doubt was resolved: great Bodies of the Enemies horse were discerned on the top of the hill on this side of Harborough, which increasing more and more in our view, begat a confidence in the General, and the residue of the Officers that he meant not to draw away, as some imagined, but that he was putting his Army in Order, either to receive us, or to come to us, to engage us

upon the ground we stood. Whilst the General was thus observing the countenance of the Enemy, directions were given to put the Army into such a posture, as that if the Enemy came on, we might take advantage of our ground, and be in readiness to receive him; or if not, that we might advance towards him.[283]

By dawn, the New Model Army was in rendezvous just to the north-east of Naseby. Being a Saturday in this growing village it was Market Day, but the wives of Naseby would have no market this day, for their community was to host one of the most famous battles in its nation's history.[284]

NOTES

[255] John Rushworth, *Historical collections of private passages of state* (7 vols., 1659-1701) part IV, vol. 1, p.36.

[256] Ibid. Rushworth inadvertently dates this occurrence as 2 July 1645, but Will Legge the Governor of Oxford certainly gave the order on 1 June.

[257] Ibid.; Sprigg, *Anglia rediviva, England's recovery* (1647),part 1, pp.26-27.

[258] Ibid.

[259] Rushworth, *Historical collections*, part IV, vol. 1, p.37.

[260] In 1642, Prince Rupert had been repulsed by the Northampton garrison, and although the King's army was far larger in 1645 the town was in good repair, thus making a siege difficult.

[261] Rushworth, *Historical collections*, part IV, vol. 1, p.37.

[262] The addition of Rawlins to Pye's regiment highlights that the list made by Joshua Sprigg, and printed in *Anglia rediviva*, was drawn up after June 1645.

[263] Rushworth, *Historical collections*, part IV, vol. 1, p.39.

[264] *The letter books, 1644-45, of Sir Samuel Luke, Parliamentary General of Newport Pagnell*, ed. H. G. Tibbut (Publications of the Bedfordshire Historical Record Society 42; HMC joint publications, 4, 1963) p.558.

[265] Sprigg, *Anglia rediviva*, part 1, p.28.

[266] Rushworth, *Historical collections*, part IV, vol. 1, p.39.

[267] Sprigg, *Anglia rediviva*, part 1, p.28.

[268] Ibid. p.29.

[269] Rushworth, *Historical collections*, part IV, vol. 1, p.39.

[270] Ibid.

[271] The ordinance made no provision for MP's to relinquish the mandate given by the electorate. The provision was solely written for members to leave the military.

[272] Sprigg, *Anglia rediviva*, part 1, p.30. The correct chronological order for the Mayors of Northampton have often been misquoted due to them taking office on 28 September

each year during this period. This made Joseph Sergeant and not Samuel Martin mayor in June 1645. I am indebted to Miss Rachael Watson, Northants Record Office, for this information.

[273] Rushworth, *Historical collections*, part IV, vol. 1, p.40.

[274] Sprigg, *Anglia rediviva*, part 1, pp.29-30.

[275] Fawsley was the ancestral home of the pro-Parliamentarian Knightley family, and had seen great meetings in the 1620-35 period of the Providence and Sayebrook Colonizing Companies.

[276] Rushworth, *Historical collections*, part IV, vol. 1, p. 41; Sprigg, *Anglia rediviva*, part 1, p.33.

[277] Arthur P. White, *The Story of Northampton* (Northampton, 1914) p.124.

[278] Sprigg, *Anglia rediviva*, part 1, p.31.

[279] Ibid. p.32; Rushworth, *Historical collections*, part IV, vol. 1, p.41.

[280] The army must have been stretched out over many miles. Gilling is a small Northamptonshire village not to be confused with Guilsborough.

[281] Rushworth, *Historical collections*, part IV, vol. 1, p.41; Sprigg, *Anglia rediviva*, part 1, p.33.

[282] Clarendon, Edward Earl of, *The history of the rebellion and civil wars in England*, ed. W. D. Macray, 6 vols. (Oxford, 1888) book IX, vol. 4, pp.37-8.

[283] Sprigg, *Anglia rediviva*, part 1.

[284] Barry Denton, *Naseby Fight* (Second edition, Leigh-on-Sea, 1991).

CONCLUSION

As the smoke cleared from the battlefield of Naseby on 14 June 1645, the picture was of complete ruin for the King. His main field army was so badly defeated, that utter destruction of the royalist cause was now a realistic possibility for the first time in three years. The New Model Army although pressed hard by Rupert and Astley's battle plan, had turned a looming defeat into a resounding victory. The New Model had in one day reduced the King's fortunes by a far greater degree than the old armies had achieved over the three preceding years. But was it the logistical superiority of the army that gave this great victory, or did luck play an important part?

The battle was in direct contrast to the fiasco witnessed at Newbury only eight months earlier. After Fairfax took it upon himself to re-assemble the army, and close with the King, the emphasis was on controlling the pace of the war, whereas at Newbury the discourse between the commanders had seen the complete opposite. It is interesting that Sir Thomas Fairfax appears to have decided ten days before the battle of Naseby, to fight an attacking war, rather than occupy defensive positions. This was totally opposite to the almost apologetic stance of the old armies. The reason could easily lie in Fairfax's early experience of the war in the Northern Association, where the enemy had primarily been the Marquis of Newcastle, and of course at Marston Moor the joint forces of the northern royalists and Prince Rupert. Naseby was the first time that Fairfax had faced the King in the field since the demonstration of Parliamentary loyalty at Heyworth Moor. This meant that Fairfax's war had hitherto been fought against local peers and foreign Princes in rebellion against Parliament, the niceties of the war encroaching upon the power balance between Royalist Oxford, and Parliamentarian London, had been at such a distance that he could perhaps engage in battle with less psychological trauma than had Essex or Manchester.

At the same time the debates and votes taken in preparation of his commission, had given to Fairfax, and through him the army as a whole, freedom from the restrictions of 1644 laid down in the Covenant, and it is clear that above all other considerations this act was the watershed in the whole course of the war. Had the New Model at Naseby faced the King under the political restrictions in place at the time of Newbury where he had been protected by the Covenant, then Fairfax like Manchester would have perhaps been constrained in his pursuance of military aims. However, it had been a conscious decision by Fairfax to refuse service if not permitted to fight without the hypocrisy which said he was preserving his majesty's person. Manchester and Essex had soothed their hearts with this former stance, but Fairfax could not. It is worthy of consideration that the removal of the words 'For King and Kingdom' had freed the New Model to prosecute the victory at Naseby to its fullest extent. At Naseby for example the New Model had not only fought upon the field of battle, but had followed up the victory to the gates of Leicester. No soldiers of the King had been allowed unmolested withdrawal from the field, there had been none of the almost farcical backing off to meet at another time, another place. This follow-up action in physically prosecuting the victory was a new aspect, a more professional method of military utilisation in the field. It is clear that had the King been within the immediate area of conflict, he was, according to the votes of March and April 1645, a legitimate target. This single hypothesis pressed by the more hawkish members in 1644, and the Independents in 1645 before its adoption by Fairfax, had emancipated the army, giving it the authority to fight to win rather than to simply protect the rights of Parliament against a faction supporting the King. The failure at Uxbridge, with Parliament being forced into a more independent stance in their negotiations, had directly influenced the way the moulding of the army had eventually materialized.

So too had the appointment of Cromwell as a temporary Lieutenant General. His inclusion in the army meant that the best aspects of the old Eastern Association horse, with its hawkish undertones, remained largely unaltered. It has been argued for three centuries that Oliver Cromwell had engineered his place in the new army, the right or wrong of that being lost somewhat in the same debate. Yet with Cromwell in the ranks, his own adherents in the army were certainly given a stronger power base from which to operate. Of the final draft of officers commissioned into the new army, six of the ten Colonels of horse were loyal to Cromwell, three regiments being commanded (if one includes John Gladman who commanded Fairfax's horse as Captain- Lieutenant) by old members of his former regiment. With such an almost clannish allegiance between the Lieutenant General and his senior officers, it cannot truthfully be argued that Cromwell did not step into a position of power, and that his presence did not influence the course of the war during 1645-46. Of course the role and temperament of Cromwell in the army had a direct bearing upon the military fortunes and radicalisation of the horse, to maintain otherwise would be acting contradictory to all known facts. It is impossible to overestimate the part the Cromwellian group played in the decision-making. Officers like Henry Ireton and Edward Whalley would, under his favour, develop their ambition into a basis for army and national influence, until the nature of their faction of the 1645 army became the standard, or status quo, for the whole military establishment.

In the Introduction to this volume, it was emphasised that at least in part this work was to examine the hypothesis set by Mark Kishlansky that the New Model Army was not the radical formation historians had maintained. The Lords, Commons and Committee of Both Kingdoms still gave power to the army, still arranged the physical logistical support, and relayed information via garrison and field commanders, but the criteria by which the three government bodies carried out their duties had seen considerable change. Likewise, in respect of its physical organisation, the army saw little change in its build up between the system used by the Earl of Essex in 1642-43 and Fairfax in 1645. In theory pay and supply of provisions were put on a more regular basis, but in practice this was only partly successful and broke down completely by the winter of 1646 through political indifference. No, where this army differed from its predecessors was in the political changes brought about between 1 and 3 April. It had been the acceptance of Self Denial, the transition of power from Lords to Commons, and the removal of the Covenant restrictions in Fairfax's commission which were fundamental to the army's fortunes. In the opinion of the present writer, this change between the power of Lords and Commons was the real change which brought victory within the year. The fears of Essex, that by undermining the old estates of Lords and King would swing the balance of power away from the established order into the hands of the middling and lower sorts of people, would through the deeds of this army transpire between 1645-46. The subsequent attempt by the Manchester group in the Lords, and the Hollis - Stapleton party in the Commons in 1646-47 to swing the power base back to its original starting position, would create the crisis of April - October in the latter year. Yet the conditions that led so undeniably to the power struggle of 1647 were framed in this moulding period in the army. It is inevitable that conditions set to fulfil one set of criteria, might, with due hindsight, perpetuate a second more radical one. In this effect, the release of the class-ridden army of 1642-45 from the command of military peers allowed Independents like Ireton to gain a foothold in politics through the army. Likewise even more radical groups, later called Levellers and Diggers, were able to attain to political status through their assimilation to the army's cause, and naturally their own grievances in the old army. It is well to remember that John Lilburne had himself left military service under the Earl of Manchester through his conscience refusing to take the Covenant, yet two years later would frame tracts of liberty for representatives of his views in the new army. This in itself suggests that the army had found a greater degree of self-expression within its makeup, obviously absent or restricted by the old restraints of the peerage commanding in the old forces. The fact that soldiers of the New Model could talk openly of desires to 'decol' or behead the King as early as 1646, and in the case of William Rainsborough, depict the severed head upon his standard, indicates the radicalism which surfaced within twelve months from Naseby.

It is clear that the power to restrain the tempers of the more politically minded members of the army was less in Fairfax's army than had been the case under Essex or Manchester. For example there was no power for Fairfax to arrest a soldier for his sectarian views, if William Packer had espoused the same views in 1645 for which a year earlier Crawford had arrested him, the charge would have been void without the intervention of Oliver Cromwell. Only acts contrary to military law were punishable, a fact that military executions were never ordered in the new army for other reason than mutiny, which was, you should note, a military crime subject to court martial, is supremely evident. Therefore we can assess without too much doubt, that while the soldiers of 1645 were the same soldiers from 1644, their attitudes were changing, and in this new army, allowed to mature. The soldier who in 1647 wrote that he was 'no mere mercenary brought together by arbitrary power' was speaking of the moulding, not the reaction to an event. He was declaring that the New Model was a free army, perhaps subject to military law, but free and open of conscience for all that.

With this in mind, was the success of the New Model Army in this moulding? Did the soldiers feel that for the first time they had a leadership that wanted to win? The answer to both these questions appears to be in the affirmative. The moulding was certainly successful with the logistical and administrative superiority leading to victory in the field and the destruction of the royalist garrison structure. It is likewise evident from the ferocity by which many royalist garrisons were stormed after 14 June 1645, that the New Model prosecuted a warlike approach towards bringing the war to its conclusion. The soldiers therefore could identify with the aims of their service, and the officers who commanded them. At no time after the assembly of the army in June 1645, did the soldiers mutiny during active service, whereas this had been a regular occurrence in previous Parliamentary armies. When elements in the New Model did mutiny in November 1647, it was largely as a result of outside political encouragement, and was quickly subdued by the Generals. The strength of the soldiers' during the crisis that engulfed the army at that time was as much due to their own discipline not to 'divide' from their inbred brotherhood, to any outside influence organising revolt within their ranks. This principle, emanating from the experience of service, was a direct consequence of the moulding bringing together a force aimed at victory, and this victory coming through their own endeavours.

The New Model Army was like no army before it in English history: it had been moulded by Lords and Commons, but commanded solely by commoners. The strength of its success lay in this single fact, and would make the army a force in British politics for some fifteen years. No King would, or could, take the throne without consent of the New Model, for this army was not bound to preserve monarchy. Indeed the King himself, when all his military fortunes had gone, surrendered not to Fairfax or the English forces, but to the Covenant Scots who were still operating under the strict Solemn League and Covenant which had been his protection at Newbury. It is plain that Charles had grasped the change in his England that a year earlier Essex had feared, and Manchester and the moderate Lords had tried to preserve. By surrendering his royal body to the Scots, the King had entered into that sanctuary where he was protected in body and state. In April

1645, the New Model had marched into the field freed of its bonds; this was the heritage of the moulding.

APPENDIX 1
THE OFFICER LISTS

The lists outlining the company and troop officers of the New Model Army, come from two main sources. The first, highly speculative list, is the army as it stood on 18 March and is detailed in both the *Journal of the House of Lords*, and John Rushworth's *Historical Collections*. The second list, represents the army as it was in the field, and was reconstituted by Rushworth and Joshua Sprigg, and printed in the latter's *Anglia Rediviva; England's Recovery*.

The list of 18 March still contains Scottish officers, and those removed for other unspecified reasons, it also includes officers who were transferred to other regiments.

The list comprised by Sprigg was the officer corps after the removal of Scots, and the changes brought about by this reorganisation.

To fully appreciate these two separate listings, and consequently assess the transition between March and May 1645, it is necessary to lay out the two side-by-side, officer by officer. It seems unlikely, as some historians have done, that the position in the list for Captains, represents in these the Captains Company number. For example in these lists, at least, it is unsafe to maintain that Richard Hill was the 6th Captain to Holborne in March, but 2nd Captain to Sir Hardress Waller in May.

The Horse were even more distinctive in their organisation, bearing no troop number, nor regimental number. It is totally inaccurate therefore, to refer to the 3rd horse, 7th foot etc, for this method belongs to a later age. The correct referencing being by Colonel's name, followed by the Company or Troop commander.

The two lists therefore look as follows

MARCH	**MAY**
THE HORSE	
Sir Thomas Fairfax	Sir Thomas Fairfax
Major Desborough	Major Desborough
Captain Swallow	Captain Adam Lawrence
Captain Browne	Captain Browne
Captain Lawrence	Captain William Packer
Captain Berry	Captain James Berry
Colonel Middleton	Colonel John Butler

Cromwell's Soldiers

Major Horton	Major Thomas Horton
Captain Foley	Captain Thomas Foley
Captain Gardiner	
Captain Butler	Captain Thomas Pennyfeather
Captain Perry	Captain Walter Parry
Colonel Sheffield	Colonel Thomas Sheffield
Major Sheffield	Major Richard Fincher
Captain Evelyn	Captain Robert Robotham
Captain Rainborugh	Captain Martin
Captain Martin	Captain William Rainsborough
Captain Robotham	Captain Arthur Evelyn
Colonel Fleetwood	Colonel Charles Fleetwood
Major Harrison	Major Thomas Harrison
Captain Fincher	Captain Coleman
Captain Lehunt	Captain Selby
Captain Coleman	Captain Richard Sanchy
Captain Selby	Captain Howard
Colonel Rossiter	Colonel Edward Rossiter
Major Twistleton	Major Thomas Harrison
Captain Markhams	Captain Coleman
Captain Nelthorpp	Captain Selby
Captain Bushey	Captain Richard Sanchy
Captain Peart	Captain Howard
Colonel Vermuyden	Colonel Bartholomew Vermuyden
Major Huntingdon	Major Robert Huntingdon
Captain Jenkins	Captain John Jenkins
Captain Bush	Captain Henry Middleton
Captain John Reynolds	Captain John Reynolds
Captain Middleton	Captain Bush
Colonel Algernon Sidney	Colonel Nathaniel Rich
Major Alford	Major John Alford
Captain Dendy	Captain Jonas Nevil
Captain Nevell	Captain Thomas Ireton
Captain Thomas Ireton	Captain Denby
Captain Bough	Captain Bough

The Officer Lists

Colonel Sir Robert Pye
Major Matthew Tomlinson
Captain Ralph Knight
Captain Henry Ireton
Captain Ralph Margery

Colonel Whalley
Major Bethell
Captain Porter
Captain Grove
Captain Horsman
Captain Packer

Colonel Graves
Major Scroope
Major General Skippon
Captain Chute
Captain Doyley
Captain Fleming

Colonel Sir M. Livesey
Major Sedascue
Captain Gibbons
Captain Hoskins
Captain Pennyfeather
Captain Barry

Colonel Sir Robert Pye
Major Matthew Tomlinson
Captain Ralph Margery
Captain Barry
Captain Thomas Rawlins

Colonel Edward Whalley
Major Christopher Bethell
Captain Robert Swallow
Captain John Groves
Captain Henry Cannon
Captain William Evanson

Colonel Richard Graves
Major Adrian Scroope
Captain Christopher Fleming
Captain William Lord Caulfield
Captain Nicholas Bragge
Captain Nathaniel Barton

Colonel Henry Ireton
Major George Sedascue
Captain Robert Gibbons
Captain William Gwilliams
Captain John Hoskins
Captain Bury

FOR THE FOOT

Sir Thomas Fairfax
Lieutenant Colonel Jackson
Major Cooke Senior
Captain Cooke Junior
Captain Beaumont
Captain Muskett
Captain Boyce
Captain Gooday
Captain Johnston

Major General Skippon

Sir Thomas Fairfax
Lieutenant Colonel Jackson
Major Cooke Senior
Captain Samuel Gooday
Captain Vincent Boyce
Captain Fulke Muskett
Captain Maneste
Captain William Bland
Captain Lea
Captain Thomas Highfield

Major General Philip Skippon

159

Lieutenant Colonel Frances	Lieutenant Colonel John Frances
Major Ashfield	Major Ashfield
Captain Samuel Clarke	Captain Samuel Clarke
Captain Streaton	
Captain Harrison	Captain James Harrison
Captain John Clarke	Captain John Clarke
Captain Bowen	Captain Bowen
Captain Gibbon	Captain Gibbon
Captain Cobbett	Captain Cobbett
Colonel Holborne	Colonel Sir Hardress Waller
Lieutenant Colonel Cottesworth	Lieutenant Colonel Ralph Cottesworth
Major Smith	Major Thomas Smith
Captain Cannon	Captain Howard
Captain Gorges	Captain Richard Hill
Captain Holden	Captain Gorges
Captain Wade	Captain John Clarke
Captain Gorges	Captain Thomas
Captain Hill	Captain Hodden
Captain Blackmore	Captain John Wade
Colonel Crawford	Colonel Robert Hammond
Lieutenant Colonel Ewre	Lieutenant Colonel Isaac Ewre
Major Saunders	Major Robert Saunders
Captain Eaton	Captain Disney
Captain Smith	Captain Ohara
Captain Ohara	Captain Smith
Captain Harvey	Captain John Boyce
Captain Disney	Captain John Puckle
Captain John Boyce	Captain Stratton
Captain John Binckle	Captain Rolfe
Colonel Berkeley	Colonel Edward Harley
Lieutenant Colonel Emins	Lieutenant Colonel Thomas Pride
Major Cowell	Major Cowell
Captain Goffe	Captain William Goffe
Captain Gregson	Captain Gregson
Captain Ramsey	Captain Sampson
Captain Jameson	Captain Hinder
Captain Leete	Captain Forgison
Captain Goddard	Captain Mason

The Officer Lists

Captain Blagrave   Captain Lago

Colonel Montague   Colonel Edward Montague
Lieutenant Colonel Grimes   Lieutenant Colonel Mark Grimes
Major Kelsey
Captain Rogers   Captain Frances Blethen
Captain Bisser   Captain Lawrence Nunney
Captain Blethen   Captain John Biscoe
Captain Nunney   Captain Wroth Rogers
Captain Wilkes   Captain William Wilks
Captain Saunders   Captain Thomas Disney
Captain Thomas Disney   Captain Giles Sanders

Colonel Aldrishe   Colonel Walter Lloyd
Lieutenant Colonel Floyd   Lieutenant Colonel Gray
Major Read   Major Read
Captain Wilkes   Captain Wilks
Captain Melvin   Captain P. Gittings
Captain Spooner   Captain Benjamin Wigfal
Captain Smith   Captain Melvin
Captain Wigfall   Captain Spooner
Captain Gettings   Captain Short
Captain Lunds

Colonel Pickering   Colonel John Pickering
Lieutenant Colonel Hewson   Lieutenant Colonel John Hewson
Major Tubbs   Major John Tubbs
Captain Axtell   Captain Daniel Axtell
Captain Husbands   Captain Azariah Husbands
Captain Jenkins   Captain John Jenkins
Captain Silverwood   Captain John Carter
Captain Carter   Captain John Silverwood
Captain Gayle   Captain Reynold Gayle
Captain Price   Captain Thomas Price

Colonel Fortescue   Colonel Richard Fortescue
Lieutenant Colonel Bulstrode   Lieutenant Colonel Jeffrey Richbell
Major Richbell   Major Thomas Jennings
Captain Dursey   Captain Edward Gettings
Captain Gettings   Captain Humphrey Fownes
Captain Fownes   Captain Young

Cromwell's Soldiers

    Captain Gimmings   Captain Gollidge
    Captain Young   Captain Whitton
    Captain Yelledge   Captain Bushell
    Captain Cobbett

Colonel Ingoldsby
Lieutenant Colonel Farrington   Lieutenant Colonel Robert Farrington
Major Cromwell   Major Philip Cromwell
Captain Duckett   Captain Henry Ingoldsby
Captain Ingoldsby   Captain Gibson
Captain Gibson   Captain Allen
Captain Allen   Captain Ward
Captain Ward   Captain Mills
Captain Mills   Captain Bamfield
Captain Bamfield   Captain Brimes

Colonel Rainsborough   Colonel Thomas Rainsborough
Lieutenant Colonel Owen   Lieutenant Colonel Bowen
Major Donne   Major Donne
Captain Horsey   Captain Crosse
Captain Westome   Captain Edwards
Captain Barber   Captain Drury
Captain Crosse   Captain ThomasDancer
Captain Edwards   Captain Creamer
Captain Lingwood   Captain Sterne
Captain Snelling

Colonel Weldon   Colonel Ralph Weldon
 This regiment fully   Lieutenant Colonel Nicholas Kempson
passed as it came up   Major William Masters
from the House of   Captain Christopher Peckham
Commons.   Captain James Fenton
                                    Captain John Franklin
                                    Captain Francis Forman
                                    Captain Jeremy Tolhurst
                                    Captain Munday
                                    Captain Kaine

THE DRAGOONS
    Colonel John Okey   Colonel John Okey
    Major Gwilliams   Major Nicholas Moore

Captain Farmer   Captain John Farmer
Captain Butler   Captain Charles Mercer
Captain Mercer   Captain Daniel Abbots
Captain Abbotts   Captain Ralph Farre
Captain Larken   Captain Tobias Bridges
Captain Farre   Captain Edward Wogan
Captain Bulkham
Captain Bridge   Captain Turpin

The two lists are now relatively simple to compare, the first examples to examine being the actual changes of officers:

THE HORSE

OMITTED   ### ADDED

Captain Robert Swallow   Captain William Packer
Colonel Middleton   Colonel John Butler
Captain Butler   Captain Thomas Pennyfeather
Colonel James Sheffield   Colonel Thomas Sheffield
                             Major Richard Fincher
Captain Fincher   Captain Richard Sanchy
Captain Lehunt   Captain Howard
Captain Bushey   Captain Henry Markham
Colonel Algernon Sidney   Colonel Nathaniel Rich
Captain Henry Ireton   Captain Barry
Captain Thomas Rawlins
Captain Porter   Captain Robert Swallow
Captain Horsman   Captain Henry Cannon
Captain William Packer   Captain William Evanson
Major General Skippon   Captain Christopher Fleming
Captain Chute   Captain William Lord Caulfield
Captain Charles Doyley   Captain Nicholas Bragge
                             Captain Nathaniel Barton
Colonel Sir M. Livesey   Colonel Henry Ireton
Captain Thomas Pennyfeather Captain William Gwilliams

This shows the actual limited changes to the men chosen to serve as officers of horse. Only eight Captains being removed, and even this figure includes the honorary troop of Skippon who served in the army in his natural role as Sergeant

Major-General. Therefore seven Captains and four Colonels were changed between the lists, from a total of 67 troop commanders, or less than one sixth of the total. Nine Captains were subsequently added to bring the horse up to strength.

It is significant that Henry Ireton was transferred from Sir Robert Pye's regiment to command that vacated by Sir Michael Livesey, after Ireton and his troop had been sent with Cromwell into the west in February. This would suggest that Ireton owed his command to Cromwell, and that he received his commission no earlier than Vermuyden's return to Fairfax's army on 8 June.

It is also likely that the Captain Butler who in the Lords' list was to serve under Middleton, was the son of John Butler who eventually commanded the regiment following rapid promotion and transfer from Okey's dragoons. The reason for this hypothesis is that John Butler had been a Captain to Hesilrige and reportedly to Waller, with his son as lieutenant to his father's troop, it is unlikely therefore that Butler Senior would have leapfrogged Thomas Horton as Major unless from an indirect line via Okey. Butler Junior eventually lost his proposed troop to Thomas Pennyfeather thus preserving the regiment intact from Hesilrige's command.

The regiment of Graves was obviously re-commissioned after the mutinies earlier in the year, removing the troop of Philip Skippon and returning the Lifeguard troop of Charles Doyley to its former position.

The changes in the Foot regiments were more numerous:

THE FOOT

OMITTED	ADDED
Captain Beaumont	Captain Maneste
Captain Johnston	Captain Lea
	Captain William Bland
	Captain Thomas Highfield
Colonel Holborne	Colonel Sir Hardress Waller
Captain Cannon	Captain Howard
Captain Gorges	Captain Thomas
Captain Blackmore	Captain John Clarke
Colonel Crawford	Colonel Robert Hammond
Captain Eaton	Captain Stratton
Captain Harvey	Captain Rolfe
Colonel Berkeley	Colonel Edward Harley
Lieutenant Colonel Emins	Lieutenant Colonel Thomas Pride
Captain Ramsey	Captain Sampson

The Officer Lists

Captain Jameson
Captain Leete   Captain Forgison
Captain Goddard   Captain Mason
Captain Blagrave   Captain Lago

Colonel Aldriche   Colonel Walter Lloyd
Captain Smith   Captain Short
Captain Lunds

Lieutenant Colonel Bulstrode   Major Thomas Jennings
Captain Dursey   Captain Whitton
Captain Gimmings   Captain Bushell
Captain Cobbett

Captain Duckett   Captain Grimes

Captain Horsey   Captain Thomas Dancer
Captain Westome   Captain Drury
Captain Barber   Captain Creamer
Captain Lingwood   Captain Sterne
Captain Snelling

Twenty-three of the Captains were therefore put out of the lists of 18 March, with resignations from four Colonels and two Lieutenant Colonels. This gives a total of twenty-nine changes of command between the two lists, out of a total of 120 company commanders, or to put it simply, almost a quarter of those originally listed. None of those omitted were transferred to other regiments.

THE DRAGOONS

OMITTED   **ADDED**

Major Gwilliams   Major Nicholas Moore
Captain Butler   Captain Edward Wogan
Captain Larken   Captain Harold Skirmager
Captain Bulkham   Captain Turpin

Both Gwilliams and Butler transferred to the Horse, meaning only two company commanders of Dragoons were omitted between lists, or exactly one fifth. The logistical changes through omissions therefore were as follows:

Cromwell's Soldiers

Colonels of Foot	4
Colonels of Horse	4
Lieutenant Colonels of Foot	2
Captains of Foot	23
Captains of Horse	6
Captains of Dragoons	2
TOTAL	**41**

Other changes were forced upon the army after marching through losses from Officers killed, death from disease, and wounds causing either permanent or temporary change in command. The following list examines this from the army's marching to the end of the first civil war.

OFFICER	CAUSE OF DEATH[1]	WOUNDED	REGIMENT[2]
Jenkins	K. Farringdon		Pickering F
Flemming		Boarstall House	Staff
Francis	K. Naseby		Skippon F
Skippon		Naseby	Skippon F
Selby	K. Naseby		Fleetwood H
Hoskins	K. Naseby		Ireton H
Bush	K. Naseby		Vermuyden H
Butler		Naseby	Butler H
Horton		Naseby	Butler H
Parry	D. 1646		
Ireton		Naseby	Ireton H
Pickering	D. Exeter		Pickering F
Tompkins	K. Naseby		Pickering F
Bethel	MW. Bristol		Whalley H
Ireton		Bristol	Rich H
Potter	MW. Naseby		Staff
Cook	K. Bristol		Fairfax F
Guilliams	K. Bristol		Ireton H
Richbell	K. Taunton		Fortescue F
Collridge	K. Taunton		Fortescue F

[1] K = Killed
  D = Died of disease
  MW = mortally wounded
[2] F = Foot
  H = Horse

Durfey	K. Bristol		Fortescue F
Fownes	K. Tiverton		Fortescue F
Ingoldsby	K. Pendennis		Fortescue F
Cobbet		Pendennis	Fortescue F
Lloyd	K. Taunton		Lloyd F
Wilkes	K. Taunton		Lloyd F
Read		Taunton	Lloyd F
Lundy		Berkeley	Lloyd/Herbert F
Melvin		Bristol	Lloyd/Herbert F
Gettings	MW. Bristol		Lloyd/Herbert F
Wigfall	K. Berkeley		Lloyd/Herbert F
Gregson		Berkeley	Harley/Pride F
Sampson		Bridgwater	Harley/Pride F
Hinder		Bristol	Harley/Pride F
Cromwell	MW. Bristol		Ingoldsby F
Ingoldsby, H.		Bristol	Ingoldsby F
Ingoldsby, T.		Bristol	Ingoldsby F
Wilkes	K. Basing		Montague F
Gayle	K. Bristol		Montague F
Donne	K. Sherborne		Rainsborough F
Crosse	K. Sherborne		Rainsborough F
Horsey	K. Sherborne		Rainsborough F
Flemming	K. Sherborne		Rainsborough F
Creamer		Sherborne	Rainsborough F
Sterne	K. Bristol		Rainsborough F
Hill	K. Bristol		Waller F
Coatesworth	K. Oxford [2nd siege]		Waller F
Francklin	K. Exeter		Weldon F
Munday	D. 1645 in the west		Weldon F

From this list it is clear that only five troop commanders were killed in the New Model Horse throughout 1645-46, with three of these being at Naseby. Therefore from 67 officers of horse, less than one in fourteen were mortalities during this period. However in the Foot twenty-six company commanders were lost from a total of 120 during the same period. This means the heavy fighting undertaken by the foot during the numerous sieges resulted in the death of just over one in five company commanders.

APPENDIX 2

THE MARCHING OF THE ARMY

In his history of the New Model, Joshua Sprigg includes a list of the marches of the army under Fairfax, but has we have seen the division of the forces means this Head Quarters listing is inadequate when following the actual movements. The first list reproduced here is the list printed by Sprigg, covering the period of 30 April to 14 June 1645, and entitled 'a journal of every dayes March of the army under the command of his Excellency Sir Thomas Fairfax; with the names of the Townes and Villages where the Head Quarters have been; the distance of miles, and how many nights the Quarters continued in each Towne or Village.'[285]

Day/month of 1645	Town/village	County	**Miles**	**No. of nights**
30 April	From Windsor to Reading	Berkshire	12	1
1 May	To Theale	Berkshire	4	1
2 May	To Newbury	Berkshire	11	2
4 May	To Andover	Wiltshire	12	1
5 May	To Salisbury	Wiltshire	15	1
6 May	To Sixpenny Hanley	Dorset	10	1
7 May	To Blandford	Dorset	7	1
8 May	To Wichampton	Dorset	7	1

The same day a Party marched Westward to relieve Taunton.

9 May	To Ringwood	Hampshire	10	1
10 May	To Rumsey	Hampshire	14	2
12 May	To Alresford	Hampshire	14	1
13 May	To Whitchurch	Hampshire	10	1
14 May	To Newbury	Berkshire	10	3
17 May	To Blewbury	Berkshire	10	2
19 May	To Newnham	Oxfordshire	9	1
20 May	To Garsington	Oxfordshire	2	2

22 May	To Marston and the siege of Oxford	Oxfordshire	4	14
5 June	To Mats Gibbon	Buckinghamshire	9	1
6 June	To Great Brickhill	Buckinghamshire	12	1
7 June	To Sherrington	Buckinghamshire	8	2
9 June	To Stony Stratford	Buckinghamshire	4	2
11 June	To Wootton	Northamptonshire	8	1
12 June	To Kislingbury	Northamptonshire	4	1
13 June	To Guilesborough	Northamptonshire	6	1
14 June	To the Battle of Naseby, and from thence to Harborough	Leicestershire	6	1

Therefore, according to Sprigg, who was listing the mileage quoted by John Rushworth, the forces with Fairfax covered 212 miles in 45 days. By modern measurements this figure appears anything up to a quarter short, for example the mileage between Whitchurch and Newbury is given as ten miles, when using the same road pattern to Fairfax the modern mileage reads thirteen miles. The reason for this difference could be explained by the use of 'old miles', which were already outdated to conform to 1,760 yards, but were still used by farmers and perhaps the military in much the same way that the kilometre has been remarkably slow to find acceptance in England. It is interesting that some English horse race courses still retain a section of track called 'the old mile' though winning posts have been set at its new mark.

The list given by Sprigg does not however deal with the known movements of Cromwell's, Weldon's and Vermuyden's small brigades. For the sake of simplicity Rossiter's regiment and those parts of Rainsborough's still in Grantham in May 1645 will not be included until joining Fairfax. The following list is pieced together from the sources referred to in the main text of this book:

Date			ARMY	
	Fairfax	Cromwell	Weldon	Vermuyden
21 April		Salisbury		
22 April		Reading		

24 April		Islip Bridge		
25 April		Bletchington		
26 April		Bletchington		
27 April		Scouting day		
28 April	Bampton, Oxon			
30 April	Windsor to	Farrington		
1 May	Reading			
2 May	Theale			
3 May	Newbury	Newbury		
5 May	Andover	Andover	Order C. B. K.	
6 May	Salisbury			
8 May	Blandford	New Bridge	PARTING	
9 May	Wichampton		Pitminster	
10 May	Ringwood	Burford	Orchard	
11 May	Rumsey		H. TAUNTON	
12 May		Woodstock	F. TAUNTON	
13 May	Alresford	Woodstock	Chard	
14 May	Whitchurh			
15 May	Newbury	Marching N. W.	Pederton	Order C. B. K.
16 May		TAUNTON		
18 May	Blewbury			Shrewsbury
20 May	Newnham			
21 May	Garsington			
23 May	Marston	Marston		Posenhall?
24 May	Wytham, OXON			Birmingham?
26 May		Wytham, OXON		
27 May		Wytham, OXON		
31 May		Ely		Melton Mowbray
1 June		Aylesbury		
3 June		Cambridge		Grantham
5 June				Market Deeping
6 June	Marsh Gibbon			
7 June	Great Brickhill	Huntingdon		Oundle
8 June	Sherrington	Cambridge		
10 June	Stony Stratford			Sherrington
12 June	Wootton			
13 June	Kislingbury	Huntingdon		
14 June	Guilesborough	Bedford		
		NASEBY FIELD		
				Guilesborough

The movement of the various elements of the army can therefore be traced for the months of April to June with some accuracy from the above column lists. For example on 5 June 1645, Fairfax was at Marsh Gibbon, while Cromwell had withdrawn to Cambridge, Weldon was trapped in Taunton and Vermuyden was riding to rejoin the main army and was quartered the night at Oundle. The movement of the forces besieging Oxford can, in some cases, be further broken down, although

Major-General Browne's forces are far from clear in either regiments or strength of number. The following list gives details of the May-June siege.

Date/month 1645	Details of military movements
17 May	Committee of Both Kingdoms order Fairfax, Cromwell and Browne to join before Oxford for siege.
19 May	New Model foot at Newnham.
22 May	Skirmish between Adjutant-General Flemming's advance party and Royalists near Marston
23 May	Breastworks thrown up on east side of River Charwell. Heavy artillery ordered up.
24 May	Boarstol House besieged by Flemming and Skippon.
25 May	Royalist garrison at Godslow deserted and later secured by Thomas Sheffield's regiment.
26 May	Cromwell's headquarters at Witcham. Fairfax at Marston. Browne at Wolvercot.
	Fairfax orders four regiments and thirteen carriages over the Charwell.
28 May	Cromwell ordered to Isle of Ely with four troops of horse.
31 May	Fairfax orders Thomas Rainsborough's regiment to besiege Gaunt House.
2 June	Skirmish at Headington.
4 June	Siege of Oxford abandoned. The temporary bridge at Islip dismantled. Boarstol House attacked by Skippon and Fairfax at night, but withstood assault.
5 June	Siege of Boarstol House abandoned.

NOTES

[285] Joshua Sprigg, *Anglia Rediviva; England's Recovery* (1647) pp.331-5.

APPENDIX 3
COATS AND COLOURS

The first reference to coat colours for the new army is given by an order from the Committee of Both Kingdoms to the Committee of Essex on 19 March. This order stated that pressed men from that county should 'be commodiously provided, as has formerly been the practice, with 1,000 red coats faced with blue.'[286] A further reference to the pressed men from the counties of Hertfordshire, Hampshire and Cambridge being added to Edward Montague's regiment, would suggest their coats were faced with blue.[287]

That the army was to be uniformly coated in red is confirmed by a reference in the newspaper *Perfect Passages*, which says that 'the men are Red-coats all the whole Army, only are distinguished by their severall facings of their coats, the Fire-locks [only] some of them are tawny coates.'[288]

It is evident that those who marched with Fairfax in April 1645 were all red-coats, except for some firelocks who came from Essex's army and were still wearing the tawny coat associated with the earl. Some of the regiments garrisoned away from the Oxfordshire, Berkshire, and London areas were possibly still wearing coats from the old armies. For example, the companies of Rainsborough's foot, which marched from Grantham, probably did not receive a new issue of coats before Naseby, whereas the three companies taken from Ayloffe's regiment did.

In April 1645, coats numbering 5,100 were ordered with white, blue, green, yellow and 1,100 orange tapes.[289] It is interesting that red coats are ordered with differing tapes, yet not linings of different colours. This could suggest that turnback facings were not being used by the New Model in 1645, but were regimented by taped edging. Alternatively, it is possible that the tapes were simply a fastening for the coat. From this information, supplemented by other documentation[290], part of the uniform details can be constructed.

REGIMENT	COAT TAPES
Fairfax	?
Skippon	?
Ingoldsby	Red[291]
Lloyd	Blue[292]
Weldon	Blue?
Pickering	Blue[293]
Hammond	White
Harley	Blue?

Rainsborough	?
Waller	Black[294]
Montague	Blue[295]
Fortescue	Green[296]

That Fairfax's foot were the 'Green regiment' is confirmed by a letter from the 'agitators' during the crisis of 1647, but it is interesting that Manchester's regiment in 1644-45 wore Green coats lined red, the complete reverse of their colours upon their reduction into Fairfax's. It would seem that Green remained the facings colour for the General from 1645 to the Restoration, when Monck's regiment, the so-called Coldstream Guards, adopted green facings to red coats.

Of the other regiments only Skippon's, Harley's and Rainsborough's remain without certain colour distinctions, although yellow and orange remain of the designated facings. Skippon had in 1642 adopted red coats lined yellow for his regiment.[297] It is not unlikely that Harley's regiment like their parental regiment, Barclay's, wore red coats with blue facings. This would give five regiments with blue, two with green, and one each with red, orange, black, white and yellow.

The coats appear to conform in the colour facing to the Ensigns, Colours or Banners of the New Model infantry. This is certainly true of Lloyd's whose banners were blue and continued a tradition from its earliest days as Lord Saye's regiment. In 1642, Saye's banners were blue with gold lions, and these continued in use until the Autumn 1644 when its field Colonel, Edward Aldriche, was awarded new banners by the Committee of Both Kingdoms. The order for Aldriche's banners from 16 December 1644 state:

> Received the day and yeare above said into his Majesties stoares within the Office of the Ordinance from Alexander Venner ye Ensignes hereafter mentioned for the supply of the Armye under the Command of his Excl the Earl of Essex. vizt -
> Silk Ensignes of blew ffllorence sarsanett with distinctions of gould coulour Laurells wth tassells to them.[298]

The order in December was for eight banners, but upon the regiments' inclusion into the New Model, two further Captains 'coulours' were ordered to bring the number to the establishment strength of ten.

The field colour for the Ensign or Banner would therefore be of the same hue as the regimental facing, with differing devices to identify the company.

NOTES

[286] *C. S. P. Dom. 1644-45*, pp.358-9.

[287] Ibid. p.460.

[288] BL Thomason Tracts E.260[32], *Perfect Passages of each dayes proceedings in Parliament*, No. 28 (April 30-May 7 1645) p.218.

[289] Stuart Peachey and Les Prince, *English Foot* (Leigh-on-Sea, 1991) p.54.

[290] C. H. Firth and G. Davies, *The regimental history of Cromwell's army* (2 vols., Oxford, 1940) vol. 1, p.323.

[291] BL Thomason Tracts, E.669[6] *French Occurences*, No. 8 (28 June – 5 July 1652) cited in Peachey and Prince, *English Foot*, p.55.

[292] Lloyd's regiment inherited blue colours from Aldriche, whose colours were made just before he relinquished his commission.

[293] BL Thomason Tracts E.304[24] *The Kingdomes Weekly Intelligencer*, No. 121 (14 Oct 1645).

[294] Ibid.

[295] *C. S. P. Dom. 1644-45*, p.460.

[296] In 1644, Fortescue's colours had been green. It is therefore likely that they adopted this colour for their facings and banners in the New Model.

[297] Stuart Peachey and Alan Turton, *Old Robin's Foot: the equipping and campaigns of Essex's infantry 1642-1645* (Leigh-on-Sea, 1987) p.33.

[298] National Army Museum, 6008/215. The contract for these same banners is included in the London Museum MS 46-78/709, and shows that each new banner cost £2 and 5 shillings.

APPENDIX 4
A TRACT FROM TAUNTON

The accounts of life in the New Model Army are many, but one of the earliest comes supposedly from an Officer of Foot in the brigade sent into the West under Ralph Weldon.

A NARRATION
OF THE
EXPEDITION
TO
TAUNTON
The Raising the Siege before it,
and the Condition of our Forces,
and the Enemies, at this present
in the WEST.
Sent from a Commander in the Army
and dated at *Chard*, *May* 18 1645

GOOD FRIEND, would I could perform my promise, with as much profit to you, as delight to my self: But in brief, Since my Apologie would but stigmatize as much your candid construction, as be a further declaration of my own weaknesse, you shall understand by these few lines our progresse from *Stanes*, our first advance on the last of *April*, by severall stages, untill we came to our generall Rendezvous at *Newbery-Wash*, *May* the 4. From whence, by twelve or one of the clock, we advanced toward *Andover*, and in the Villages adjacent, with seven Regiments of Foot, to the number, I suppose, of 10 or 12000 Foot; for the Horse most of them not then as yet come in: I would write no more then what I am able to testifie, either by my own, or from the hands of good Authors.

From *Andover* on *May* the fifth, we drew up our severall Regiments a mile from the Town and staid two or three hours, called a Councell of War, where were cast five or six; one a Renegado, and four more, authors of the mutiny in *Kent*, who cast dice for their lives; one of them, the Renegado a Parsons son, were executed in a Village on a Tree in the High-way, in *Terrorem*; the Parsons son, as was said, in the same Town where he was born, both of them died as they lived, like Sotts: But how the great Judge past his sentence on them, I have not to say. Next day, *May* the sixth, was Proclamation made, *That it should be death for any man to plunder*, at which, our old Horse-Dragoons, somewhat guilty, made answer, *If the Parliament would pay truely, let them hang duely*: Which hath caused so much good order in our march, that to my best inquiry, I have not heard of any man to complain to loose an Ox, Sheep, Lamb, Hen, no, nor an Egg, save in our hard march, hot dayes, vacancy of Townes, or Houses, over the Plain, made them inordinately desire drink, or covet for water in the Villages we past. To give you the particular of our severall Stages, would be as difficult, as needlesse: Since for the most part we took Barnes, and Hedges for our nights repose, after our hard and hot-dayes marches, untill within the compasse of eight dayes. We came on the second of this instant, within the sight of *Taunton*, where upon the Hills, when we came within ten miles of it, having advantage of ground, we gave them a peal of our Artillary, ten of our peeces being discharged to give them notice of our approach, which yet did them no good, by reason, that on the Thursday before, which was the eighth of this instant Month, the enemy drew out a party of their Horse and Foot, with some peeces of Cannon, and skirmisht in sight of the Town, but only with Powder; in sine, they made the Town beleeve, that *Fairfax* who was coming to relieve them, was there beaten, hoping by this stratagem, to have cut them off by an Ambuscado: But God withheld them, they kept close to their works, and when the enemy returned, they fell to firing of the Town, and told them, you Roundheaded Rogues, you look for relief, but we have relieved them, and *Goring* is coming on, and we will not leave one House standing, if you will not yeild; then they played with their Granadoes and Morter peeces so hot, and so long, that they fired the Town: So that, I beleeve, the one half of the Town, which was two long streets of the Subburbs, be both burnt down to the ground; and the mean while, they stormed most furiously, but they met with a Gallant Commander in chief, Colonell *Blake*, and his stout Souldiers, that gave them such showers of lead, that from good hands it is reported, 1200 at least, there sacrificed their filthy lives, and left their carkasses: The Town is all, from the beginning, to the raising of the siege, hath lost 200 men. On Friday they had work enough to bury their dead, and bethink themselves of saving their living. *Hopton* sent a parly to resign the Town upon conditions, *Blake* returned him answer, he had four pair of Boots yet left, and he would eat three pair of them, before he should have it.

On Saturday we came to *Chard*, within eight miles of it; on the Lords day, orders were given to beat our Drums by day light, and accordingly, drew up our whole Army, Horse and Foot; and all-though by command from the Parliament, our noble Generall *Fairfax*, Generall-Major *Skippon*, and that Regiment which formerly was *Barkleys*, were commanded back when we were at *Blandford*, which was a sad breakfast to most, both Officers and Souldiers, were after sad salutes, and watery eyes, like the parting of Husbands from their Wives, and dear friends, yet we were thinking of nothing more then resolution to obey, and action to perform, that great work for the which we were sent, with four Regiments of Foot, to wit, Colonell *Weldon*, who as eldest Colonell, a gallant, wise, and brave Gentleman,

Commander-in-chief, Colonell *Fortescu*, Colonell *Floyd*, and Colonell *Englesby* Regiments; and as we past, came in for our recrute, and met us about *Dorchester*, six Companies of the Skie-colour Regiment of Colonell *Morrells*; and on Saturday, as many Colours from *Lime*, those old, brave Blades, We had a fine Body of Horse, of some 1500 or 2000. and 4. or 5000 Foot, where I never beheld men of all sorts, of more promising courage, resolution, all as one man, sweetly combined against the common enemies of mankind, such love amongst themselves, Horse and Foot. One passage I will relate, though I hate prolixitie, a brave gallant Fellow, but a common Soldier, cries out to the Horse as they marcht by: O brave Horse, go on, shew them no more mercy, then to a Louse: Remember *Cornwall*; To whom a brave Captain of the *Plimmouth* Troop replied, O Fellow Souldier, let us remember of God, and not fight in malice, but do his work, and leave the successe to him, and you shall see, through Gods mercy, we will stand close to you, O you gallant Foot; but I may not be tedious, yet surely, braver courage was never seen, then even then, when a party of the enemies Horse, and ours, faced each other; our Forlorn Horses meet, and exchange some Pistols, put them quickly to the Tryall of their heels; but after we were drawn up into Battalia, expecting when to be charged, and made choice of our ground, no enemy appeared, we went on to the very Brow of the Hill in Battalia, and saw betwixt that and *Taunton*, nothing but inclosures, not minding to adventure all our Horse into a pitfold; the Lanes in many places from thence, we could not march above four or six in brest: The Agitant of the Horse surely a gallant man, a Dutchman, and some eight or nine others, fall down to *Pitminster*, and without his Dublet, only in his Shirt, incountered a Troop of the enemies Horse, being as is said, *Hoptons* Lifeguard, for they were commanded by his own Cornet, one *Brown* an *Irish* man, who suriously charges them, and cries, why do you not fire you cowardly Rogues, spent one Pistoll, charged them thorow, and killed three or four with his own hands, his Sword being all bloody up to the very Hilt; they all run as fast as they could, the Dutch Agitant wheels off, and retreats, lost not one man in the first Rout; the Cornet and some twelve more, faced about, and on them the Agitant charges a fresh, having killed two or three of them, took four of them prisoners, rides up to the Cornet, and cries quarter, he denies it, but sets Spurs to his Horse to run away, but he was soon overtaken by his Pistoll, which ended his journey by a brace of Bullets in his back, fell from his Horse, he brought him of alive, but he soon died; and being demanded why he refused quarter, made answer, He could not in honour desire it, seeing so many to be beaten, and run away from so few: But questionlesse, the man thought of his Nation, and dreading a halter, chose a more honourable death. All this while, we have not one word from the Town; whereupon, we gave them two or three peeces of Cannon, but they were cautious, and perswaded we were the enemy, who indeavoured to draw them forth; for so the enemy suggested, that Goring was come to relieve them, and that their Rear might not be discovered: toward Evening, we sent a party of Horse, who approached to their very Works, the enemy having drawn off their Guns, and their Rear upon their march, the Town never before, having any notice of our Forces, that they could confine on as friends there, about six of the clock fell out upon their Rear, killed some, and took other prisoners. We marcht with our whole Body to *Pitminster*, and then within two miles of the Town, took up our quarters in the Fields, and on Monday morning our Colonels go to *Taunton*, give order for our whole Army to retreat back to *Chard*, where we quartered on Saturday; and the fourteenth is the first dayes rest, the Army hath had, from our first dayes motion Westward; which if you consider of, is one of the greatest expeditions, and gallantest marches, that ever this unhappy War produced; if you do but waigh the length of the way, the incumbrances that attend an Army, with their Train, and Artillery; many new Souldiers, hard quarters, exceeding cold nights, and as hot dayes: Lett God have the Glory, our Colonells and Officers the praise; who of my knowledge, have marcht two or three dayes on Foot, and never took their Horse, but still in the head of their Regiments, gave good incouragement by their own examples, and then God so ordering, that not a man of us miscarried; for my own Company, I can say, not a man sick after we left *Newbery*, and few or none went from us, but all stick close, valiantly resolved to fight and die, yet the Town was relieved without the losse of any mans blood, and a terrour strook into the hearts of their enemies. We may say, God fought for us, and of him, we will make our boast all the day long; To whom be all the praise, who lives for ever and ever, *Amen, Amen.*

    Thus far have I brought you to our own Armies expedition; but now for the enemy one Word, and I have done, though neither this nor future ages will believe, nor should I my self, who have formerly known these parts, had not my own eies beheld it: To see one or two Houses ruined in a place, had been no great matter, but all the way we marcht from *Okingham* to *Taunton*; no place especially, where Religion was most eminent, but you might track the divell by his cloven Foot: Such devastation of Houses, nay, depopulations in many places; and those Fields, Pastures, Plains, formerly beautified, and inricht with Flocks, and Herds: You may passe ten miles, and scarce discern any thing; rich Pastures, but no Cattle left to eat them. You would suppose the great *Turk*, his Janisaries and Armies, rather than their Native Prince his Souldiers had been there: Who would think a King who was so tender hearted, as to charge *Hotham* so deeply in his Answer to the Parliament, declared 1642. for drowning the Meadows about *Hull*, and was formerly so carefull for the good of the Subjects disabusing by the severall Manufacturers of Sope, Cards, Dice, Pinnes and c. should now lay all desolate where he hath any footing. O that those Counties of *Kent, Essex, and c.* which complain of heavy Taxes, would compare their Estates with the forlorn West: His Majestie complains, that the Flowers of his Crown, the Prerogative Royall, should be infringed; and yet the Jewells of the ancient Crown sold to buy us such an unheard of misery, by

Walloons, Irish, French, Dutch, and c. If all this will not please the Queen, the Papists, Jesuites, yea, the divell himself; I know not how they can studie more to give man, and provoke God. Let them palliat the Prince, and tell him, he is to give an account to none but God; yet sure that will be found an hard reckoning at last: A sad maxime, that no way so sure to settle a King in his Throne, as to Pave its way thorow the blood and ruin of his people and Kingdom. We have heard of three or four Kings, in four or five yeers, and yet the Kingdom to flourish; but we never heard, that one King should destroy two or three Kingdoms, rather then suffer the least affront, or the least twing of the Toothack; and cursed be those Councels of the Rabbies in *Oxford*, and divels of *Rome*, who so perswade. If this be Regall Government, I know not what to call Tyranny; yet King *James*, though none of the best Princes, nor worst Politician, hath written *Basilicon Doron*, and shrewdly descanted on this Theam. However *Solomon* could judge of the true mother by her affection, rather to save her childes life, though she lost her childe formerly: Good Princes have been the shields and saviours, not destroyers of their Countrys. If the King should miscarry (which we abhor to think or desire) yet we may have many Kings hereafter; But if three Kingdoms perish, what is Monarchy without Majestie; and what Majestie can be upheld with beggery, misery, and slavery. If this be the Protestant Religion, its a strange one: yet so it must be called, yea, *Hopton* when he saw he could not take, yet cruelly burnt that distressed Town; and when it was all in flames, called, and pulled out by the ears, those distressed people adjacent, to look and behold the flames, with execration and scorn; yet after he had two preachments, no doubt, but to give God thanks, like the Duke of *Alva*, who before Dinner, gave a good Grace to his meat, thanking God for his buchery of so many thousands in a few yeers: This is *that Hopton* formerly accounted Religious, Honest, Noble, so degenerate by the Councell of *Tobi Matthews*, and old *Cottington*, and his Uncle, Sir *Arthur Hopton*, Spanish compliance, and all grounded upon his own beggerly estate, so pitifully torn, and out at heels, that he is become the monster of mankinde. In brief, the poor people come from all parts, rejoycing, praising God, and thanking us for delivering them from those Beasts of prey, who before this time, had no Trade, Market, Commerce, or society with each others. Now their faces begin to shut out the former wrinkles, and smilingly tell us, we have shrewdly galled the Cabballers, shower and shower, they be all ago: The Lord keep them as far as we have left them; for we have orders to march away Eastward this day, having had not one days rest this fourteen dayes, before this time: And now our men are cheerfully marching, and we hope you are praying, that you may never taste that in *London*, which we have seen in the West. O if *Kent* did know their happinesse, they would not be so mad to purchase such misery, at so dear a rate; however, they blesse themselves with hopes of their King, these poor souls have found they have King enough. Sir, I have no more to say, but desire three or four words, how things go at *Scarborough*, and in the North.

Chard, 18 of May, 1645.

FINIS

The source of the content of this tract appears twofold, the first a hand written letter from an officer of foot in Weldon's Brigade, and the second being by way of additional material supplemented to the whole to give a political bias. That the tract contains material added in London is clear, to actually write the eight pages of the printed tract, by quill, and after an exhausting march, would constitute in the region of twenty pages. The average letter written in the field appears to be no more than two pages, suggesting that additions were made later.

The regiment of the unnamed commander is unknown, yet he says 'my Company' indicating that he was an officer in a foot regiment. There are also numerous references to Kent, where indeed Weldon's own regiment was raised, but whether this fact points to the author being of that unit is conjectural only.

It is evident from the section of the tract dealing with kingship, that a section of the Army, and furthermore Parliament itself, was hardening their stance against Charles the man, writing openly of 'what is Monarchy without Majestie', and quite clearly stating that the throne is in Majestie which is not tyranny. Such a point returns to the question raised by the Covenant and suggests that the army of 1645 marched under a more radical commission than had Essex's own raw recruits.